The Mechanics of Motion
(Force, Friction and Energy Explored)

By

PRASHANT KUMAR LAL

The Mechanics of Motion
(Force, Friction and Energy Explored)

© **Copyright, 2024, Prashant Kumar Lal, Author**

All rights reserved. No part of this book may be reproduced, stored in a retrieval system, or transmitted, in any form by any means, electronic, mechanical, magnetic, optical, chemical, manual, photocopying, recording or otherwise, without the prior written consent of its writer.

ISBN: 9798301401510 Paperback
Imprint: Independently published

Published By: Prashant Kumar Lal
R – 133, Phase 2, Sarabha Nagar Extension
Pakhowal Road, Ludhiana (Pb), Pin – 142022, (INDIA)
Email: prasshantlal1961@gmail.com,
Phone: +91 9478624982

Publishing Fields: Science & Technology, Educational Books

The Mechanics of Motion
(Force, Friction and Energy Explored)

By: Prashant Kumar Lal

About The Book

"The Mechanics of Motion: Force, Friction, and Energy Explored"

Is a fascinating dive into the foundational concepts of physics, making it an essential read for enthusiasts, students, and professionals eager to deepen their understanding of motion and the forces behind it. This book blends theoretical concepts with practical applications, offering a clear and accessible explanation of how the laws of physics shape our world.

Overview of the Book

The book is structured to unravel the complexities of motion in a step-by-step manner, beginning with basic concepts like force and inertia, and gradually progressing to advanced topics like friction, energy conservation, and real-world applications of motion. Its meticulous approach ensures that readers of all backgrounds can grasp the intricacies of mechanics.

Key Themes
1. Force and Motion
The book delves into Newton's Laws of Motion, explaining their relevance to everything from a child's swing to the movement of celestial bodies. It offers examples from everyday life to elucidate the principles

of dynamics.

2. Friction: A Double-Edged Sword

Friction, often seen as a hindrance, is portrayed in a new light. The author explores its dual role in facilitating movement and causing wear and tear. Thought-provoking case studies, such as the role of friction in transportation and engineering, make this section particularly engaging.

3. Energy and Work

The book presents a seamless connection between motion and energy. Concepts like kinetic and potential energy, energy transfer, and the role of energy in sustaining life and technology are discussed with illustrative examples and experiments.

Unique Features

Hands-On Experiments: The book includes DIY experiments for readers to observe principles in action, making learning interactive and enjoyable.

Historical Insights: Anecdotes about the evolution of mechanics and contributions by pioneers like Galileo, Newton, and Einstein enrich the narrative.

Real-World Applications: Each chapter connects theory with its applications, ranging from sports to space exploration, making the concepts relatable.

Who Should Read It?

This book is perfect for:

- Students looking to strengthen their fundamentals in physics.

- Educators seeking a resource to explain complex topics in a simple and engaging way.

- Professionals and engineers who want to revisit the basics or apply mechanical concepts to practical problems.

The Mechanics of Motion: Force, Friction, and Energy Explored is more than a textbook—it's an exploration of the forces that govern our universe. Its ability to simplify complex ideas while maintaining scientific rigor makes it a standout in the realm of physics literature. Whether you're a beginner or an advanced learner, this book promises to be a valuable addition to your collection.

Would you like to explore a specific topic from this book in more detail?

About The Author

"The Mechanics of Motion: Force, Friction, and Energy Explored"

Prashant Kumar Lal is an accomplished educator, author, and consultant with a profound passion for physics and education. With over 38 years of experience in teaching, administration, and academic leadership, he has dedicated his life to shaping young minds and promoting a deeper understanding of scientific principles.

Born in Kathmandu, Nepal, and educated under the Jesuit Fathers, Mr. Lal's upbringing instilled in him a love for knowledge and discipline. He holds a strong academic background in Physics, which became the cornerstone of his illustrious career. Over the years, he has served as a teacher, a principal, and a mentor, nurturing countless students to achieve their academic and personal potential.

After retiring from his formal role as a school Principal, Mr. Lal founded Prashant Educational Consultancy Services OPC Pvt Ltd, an organization committed to supporting schools in academic planning, administrative efficiency, career counselling, and teacher training. His work continues to impact the educational sector, inspiring institutions to adapt to modern trends while maintaining a strong foundation in values and excellence.

A prolific writer, Mr. Lal has authored several books, including Image of My Experiences – A Book of Poetry, Speeches from the Desk of the Principal, and The Legend of Inara Wali. His writings often reflect his rich life experiences, integrating history, philosophy, and mythology with modern-day relevance.

In The Mechanics of Motion: Force, Friction, and Energy Explored; Mr. Lal brings his extensive teaching experience to life. His ability to simplify complex concepts, engage readers with relatable examples, and inspire curiosity makes this book an invaluable resource for students and educators alike.

When not immersed in academic pursuits, Mr. Lal enjoys music, playing the harmonium and keyboard, trekking, and spending quality time with his family. A lifelong learner and teacher, he remains committed to igniting a passion for science and education in all those he encounters.

Preface

Motion is a phenomenon that surrounds us every day, whether it's the rhythmic movement of the pendulum, the friction beneath our feet, or the immense forces propelling rockets into space. The principles governing motion are as old as the universe itself, yet they remain as captivating and relevant as ever. It is this timeless intrigue that inspired the creation of The Mechanics of Motion: Force, Friction, and Energy Explored.

This book is designed to be a journey into the heart of mechanics—a field of physics that explains the "why" and "how" of motion. With clarity and precision, it unfolds the interconnected principles of force, friction, and energy, demonstrating how they work in harmony to shape the world around us. The aim is not only to educate but also to inspire curiosity and a sense of wonder for the natural laws that govern our existence.

From Newton's laws to real-world applications, this book offers a balance of theory and practice. Whether you are a student navigating the complexities of physics, a teacher searching for ways to make concepts engaging, or simply a curious mind eager to explore, this book caters to all. Each chapter is enriched with historical anecdotes, hands-on experiments, and practical examples, making the content relatable and memorable.

As you read, you will encounter stories of the great minds who laid the foundation for modern mechanics, insights into the role of friction in technology and engineering, and a deeper understanding of energy's transformation in our daily lives. The material is structured to build progressively, ensuring that readers of all levels can follow along with ease.

The journey into mechanics is both humbling and empowering. It unveils the elegance of the universe and reminds us of the boundless potential of human ingenuity. It is my hope that this book will not only enhance your understanding of motion but also spark a lifelong passion for exploring the forces that shape our world.

Let us embark on this exploration together, guided by the eternal quest for knowledge and the beauty of discovery.

Prashant Kumar Lal

Author

I extend my deepest gratitude to those who have been my unwavering pillars of strength, encouragement, and love. Your support has been my greatest blessing, inspiring me to pursue my passions and overcome every challenge with resilience. To those who have shared in my journey, stood by me in my endeavors, and celebrated my successes, I am forever thankful. Your belief in me fuels my determination and fills my heart with gratitude beyond measure.

The Mechanics of Motion: Force, Friction
and Energy Explored   By: Prashant K

Contents

"The Mechanics of Motion: Force, Friction, and Energy Explored" sounds like a fantastic title for a physics book! Here's how you could potentially break down and structure its contents to make it engaging:

1. Introduction to Mechanics and Motion

Setting the foundation with basic concepts and terminology.
Introducing Newton's three laws of motion as the cornerstone.   Pg - 14

2. Force and Its Many Forms
Types of forces (gravitational, normal, frictional, tension, etc.).
Practical examples to show how forces interact in daily life.   Pg - 74

3. Friction: The Hidden Resistance
The role of friction in motion and how it can both help and hinder.

Types of friction: static, kinetic, rolling, and fluid.
Real-life applications and implications of friction.   Pg - 94

4. Circular Motion and Centripetal Force
Understanding rotational dynamics and centripetal force.
Examples of circular motion, like satellites, car tires, and planetary orbits. Pg - 121

5. Work, Energy, and Power
Defining work and its relationship to force and displacement.
Exploring different forms of energy: kinetic, potential, thermal, etc.
The principle of conservation of energy and real-world applications.
How energy transfers in closed systems and open systems.
Real-world examples of energy conservation, like roller coasters and pendulums.
Highlighting famous experiments or phenomena that exemplify these principles. For example, the behavior of projectiles, pendulums, and circular motions in sports.
                Pg 156

Chapter - 1

Introduction to Mechanics and Motion

Mechanics is the cornerstone of physics, focusing on how objects move and the forces that influence them. It forms the basis for understanding natural phenomena and engineered systems, from the orbit of planets to the mechanics of a car engine. In this chapter, we explore the principles of motion, the laws governing it, and its practical applications, laying the groundwork for advanced studies in physics.

What is Motion?

Motion is defined as the change in the position of an object concerning a reference point over time. For instance, consider a car moving along a road. The car's position changes relative to the buildings around it, indicating motion. Motion is everywhere: the wind blowing, the flow of water in a river, or the rotation of the Earth on its axis. Mechanics is the cornerstone of physics, focusing on how objects move and the forces that influence them. It forms the basis for understanding natural phenomena and engineered systems, from the orbit of planets to the mechanics of a car engine. In this chapter, we explore the principles of motion, the laws governing it, and its practical applications, laying the groundwork for advanced studies in physics.

Types of Motion

Motion can be classified into three primary types:

1. *Linear Motion:*

Movement in a straight line. Examples include a train moving along a track or a person walking down a hallway. Linear motion is the simplest form of motion, often described by the relationship between distance, velocity, and acceleration.

Equation: $v = d/t$, where 'v' is velocity, 'd' is distance, and 't' is time.

Linear motion is a type of motion where an object moves in a straight line. It's one of the simplest types of motion to understand and analysis.

Key Concepts:

a. Distance: The total length of the path travelled by an object.

b. Displacement: The straight – line distance between the initial and final positions of an object.

c. Speed: The rate at which an object covers distance.

d. Velocity: The rate at which an object changes its position, including direction.

e. Acceleration: The rate at which an object changes its velocity. a= dv/dt = change in velocity / Time

Types of Linear Motion:

a) Uniform Motion: An object moves with a constant speed in a straight line.

b) Non-Uniform Motion: An object moves with a varying speed in a straight line.

Applications of Linear Motion:

Linear motion is ubiquitous in our daily lives and has numerous applications in various fields:

a) Transportation: Cars, trains, and airplanes all rely on linear motion.

b) Sports: Athletes like runners and swimmers use linear motion to achieve their goals.

c) Manufacturing: Assembly lines and robotic

arms utilize linear motion for efficient production.

d) **Physics Experiments**: Many physics experiments involve linear motion, such as studying projectile motion or the laws of motion.

Real-World Examples:

a) A ball rolling down a hill: The ball's motion is linear, but its speed increases due to gravity.

b) A car moving on a straight road: The car's motion is linear, and its speed can be constant or varying.

c) A bullet fired from a gun: The bullet moves in a straight line until it is affected by gravity and air resistance.

Distance-Time Graphs: These graphs show how distance changes over time. A straight line represents constant speed, while a curved line indicates changing speed.

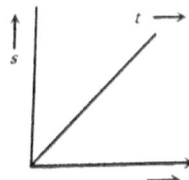

Velocity-Time Graphs: These graphs show how velocity changes over time. A horizontal line represents constant velocity, while a sloping line indicates acceleration or deceleration.

The Mechanics of Motion: Force, Friction
and Energy Explored   By: Prashant Kumar Lal

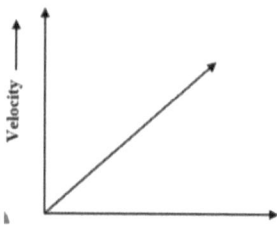

By understanding the fundamental concepts of linear motion, students can better appreciate the world around them and apply these principles to solve various problems.

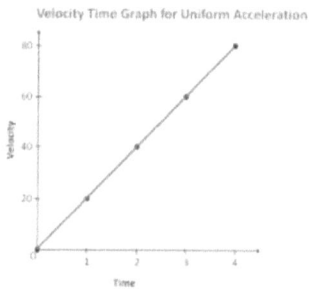

The velocity – time graph of a linear motion as shown in the figure below. The displacement & distance from the origin after 8 sec

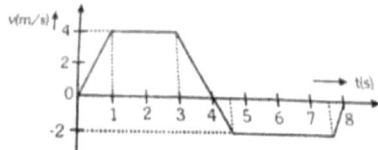

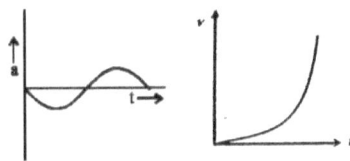

2. Circular Motion:

Motion along a circular path. The Earth revolving around the Sun is a prime example. Circular motion introduces concepts such as angular velocity and centripetal force.

Example: A stone tied to a string and swung in a circle.
Imagine you're swinging a ball on a string around your head. The ball moves in a circular path, right? That's circular motion! It's when an object moves in a circular path around a fixed point.

Key Concepts
   a) **Uniform Circular Motion**: When an object moves in a circle with a constant speed.

   b) **Centripetal Acceleration**: This is the acceleration that pulls an object towards the center of the circle. It's always perpendicular to the object's velocity.

   c) **Centripetal Force**: The force that causes centripetal acceleration. It's also directed towards the center of the circle.

Examples of Circular Motion

a) The Moon Orbiting the Earth: The Moon's path around the Earth is nearly circular.
b) A Car Turning a Corner: The car's wheels follow a curved path.
c) A Ferris Wheel: The cabins move in a circular path.
d) A Swirling Top: The top spins in a circular motion.

Visualizing Circular Motion

Here's a diagram illustrating a body in circular motion:

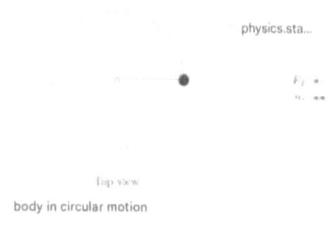

body in circular motion

Mathematical Description

a) Velocity: In circular motion, the velocity vector is always tangent to the circular path.
b) Acceleration: The acceleration vector points towards the center of the circle.
c) Centripetal Acceleration:
d) $a_c = v^2 / r$
e) $a_c = \omega^2 \times r$
f) $\omega = 2\pi / T$ (angular velocity)

g) T is the time period of one complete revolution

Derivation of Centripetal Acceleration

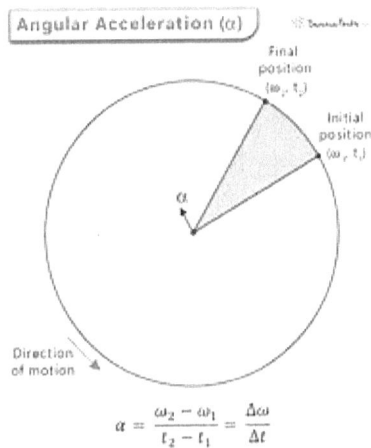

[Diagram showing a particle moving in a circle]
Consider a particle moving in a circle of radius r with a constant speed v. In a small - time interval Δt, the particle moves from point A to point B, covering a distance $\Delta s = v\Delta t$. The change in velocity, Δv, is directed towards the center of the circle.
Using geometry, we can show that:
$\Delta v / v \approx \Delta s / r$
As Δt approaches zero, this becomes:
$a_c = \lim(\Delta t \to 0) \, \Delta v / \Delta t = v^2 / r$

Centripetal Force
According to Newton's Second Law, $F = ma$. For circular motion, the force causing the centripetal acceleration is called the centripetal force.

$F_c = m \times a_c = mv^2 / r$

Applications of Circular Motion
a) Centrifuges: Used to separate substances of different densities.
b) Roller Coasters: Designed to provide thrilling experiences using circular motion.
c) Satellites: Orbit the Earth in circular paths.
d) Motors and Generators: Utilize circular motion to convert energy.

Understanding Circular Motion
Circular motion is a fundamental concept in physics, with applications in various fields. By understanding the key concepts and mathematical relationships, you can appreciate the beauty and power of this type of motion.

3. Rotational Motion:

An object rotates around its axis. For instance, a spinning top or the rotation of the Earth.
Key characteristic: Every point on the object follows a circular trajectory relative to the axis of rotation.
Imagine a spinning top or a merry-go-round. These are examples of rotational motion, a type of motion where an object moves in a circular path around a fixed axis.
What is Rotational Motion?
Rotational motion is the motion of an object around a fixed axis. This axis can be internal or external to the object. Each point on the object moves in a circle around this axis.

Key Concepts in Rotational Motion:
* Angular Displacement (θ):
  * Defined as the angle through which an object rotates.
  * Measured in radians.
  * 1 radian = 57.3 degrees.
* Angular Velocity (ω):
  * Rate of change of angular displacement.
  * Measured in radians per second (rad/s).
  * Formula: $\omega = \Delta\theta/\Delta t$
* Angular Acceleration (α):
  * Rate of change of angular velocity.
  * Measured in radians per second squared (rad/s²).
  * Formula: $\alpha = \Delta\omega/\Delta t$

Relationship between Linear and Rotational Motion:
For a point at a distance r from the axis of rotation:

a) Linear velocity (v) = ωr

b) Linear acceleration (a) = αr

Moment of Inertia (I):

a) Rotational analog of mass.

b) Measures an object's resistance to rotational motion.

c) Depends on the mass distribution and shape of the object.

d) Formula: $I = \Sigma mr^2$

Torque (τ):

a) Rotational analog of force.

b) Causes angular acceleration.

c) Formula: $\tau = I\alpha$

Newton's Second Law for Rotational Motion:

a) The angular acceleration of an object is directly proportional to the net torque applied to it and inversely proportional to its moment of inertia.

b) Formula: $\tau = I\alpha$

Conservation of Angular Momentum:

a) In the absence of external torques, the total angular momentum of a system remains constant.

b) Formula: $L = I\omega$

Applications of Rotational Motion:

a) Everyday life: Spinning wheels, ceiling fans, clocks.

b) Engineering: Rotational machines like turbines, motors, and generators.

c) Sports: Spinning balls, figure skating spins.

d) Astronomy: Planetary and stellar rotation.

Graphical Representation:

a) Angular displacement-time graph: Shows how angular displacement changes with time.

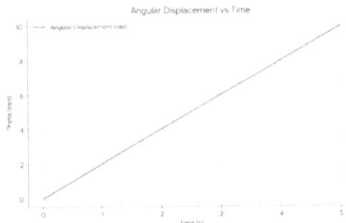

b) Angular velocity-time graph: Shows how angular velocity changes with time.

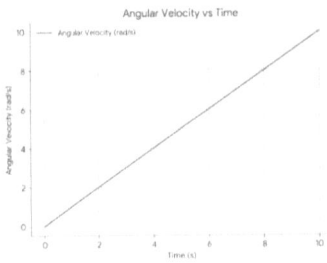

c) Angular acceleration-time graph: Shows how angular acceleration changes with time.

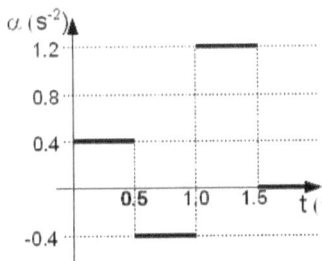

Derivations and Mathematical Operations:
* **Derivations**: Involving calculus to derive relationships between angular quantities.

Here are the derivations involving calculus for the relationships between angular quantities:

Angular Displacement and Angular Velocity

The angular displacement (θ) of an object is defined as the angle through which it rotates. The angular velocity (ω) is defined as the rate of change of angular displacement with respect to time.

Mathematically, we can express this relationship as:

$$\omega = d\theta/dt$$

where ω is the angular velocity, θ is the angular displacement, and t is time.

Angular Velocity and Angular Acceleration

The angular acceleration (α) is defined as the rate of change of angular velocity with respect to time.

Mathematically, we can express this relationship as:

$$\alpha = d\omega/dt$$

where α is the angular acceleration, ω is the angular velocity, and t is time.

Angular Displacement and Time

We can integrate the angular velocity with respect to time to get the angular displacement:

$$\theta = \int \omega \, dt$$

This equation shows that the angular displacement is equal to the integral of the angular velocity with respect to time.

Angular Velocity and Radius

For an object moving in a circular path, the angular velocity is related to the linear velocity (v) and the radius (r) of the circle by the following equation:

$\omega = v/r$

This equation shows that the angular velocity is equal to the linear velocity divided by the radius.

Angular Acceleration and Radius

For an object moving in a circular path, the angular acceleration is related to the linear acceleration (a) and the radius (r) of the circle by the following equation:

$\alpha = a/r$

This equation shows that the angular acceleration is equal to the linear acceleration divided by the radius.

Torque and Angular Acceleration

The torque (τ) is related to the angular acceleration (α) and the moment of inertia (I) by the following equation:

$\tau = I\alpha$

This equation shows that the torque is equal to the product of the moment of inertia and the angular acceleration.

Angular Momentum and Angular Velocity

The angular momentum (L) is related to the angular velocity (ω) and the moment of inertia (I) by the following equation:

$L = I\omega$

This equation shows that the angular momentum is equal to the product of the moment of inertia and the angular velocity.

These are the main relationships between angular quantities involving calculus.

* **Mathematical operations: Using trigonometry, algebra, and calculus to solve problems.**

Here are the mathematical operations for the relationships between angular quantities using trigonometry, algebra, and calculus:

Angular Displacement and Angular Velocity

$\omega = d\theta/dt$

Using trigonometry, we can express the angular displacement θ as:

$\theta = \arctan(y/x)$

where (x, y) are the coordinates of the point on the circle.

Using algebra, we can differentiate the angular displacement with respect to time to get:

$d\theta/dt = \omega = (x\, dy/dt - y\, dx/dt) / (x^2 + y^2)$

Using calculus, we can integrate the angular velocity with respect to time to get:

$\theta = \int \omega\, dt$

Angular Velocity and Angular Acceleration

$\alpha = d\omega/dt$

Using algebra, we can differentiate the angular velocity with respect to time to get:

$d\omega/dt = \alpha = d^2\theta/dt^2$

Using calculus, we can integrate the angular acceleration with respect to time to get:

$\omega = \int \alpha\, dt$

Angular Displacement and Time

$\theta = \int \omega\, dt$

Using calculus, we can evaluate the integral to get:

$\theta = \omega t + C$

where C is the constant of integration.

Angular Velocity and Radius
$\omega = v/r$
Using algebra, we can rearrange the equation to get:
$v = \omega r$

Angular Acceleration and Radius
$\alpha = a/r$
Using algebra, we can rearrange the equation to get:
$a = \alpha r$

Torque and Angular Acceleration
$\tau = I \alpha$
Using algebra, we can rearrange the equation to get:
$\alpha = \tau / I$

Angular Momentum and Angular Velocity
$L = I \omega$
Using algebra, we can rearrange the equation to get:
$\omega = L / I$

Here are some examples of mathematical operations using trigonometry, algebra, and calculus:

Example 1: Finding Angular Velocity
A wheel rotates through an angle of 30° in 2 seconds. Find its angular velocity.
$\omega = d\theta/dt = (30° / 2 \text{ s}) = 15°/s$

Example 2: Finding Angular Acceleration
A car accelerates from rest to a speed of 20 m/s in 4 seconds. Find its angular acceleration.
$\alpha = d\omega/dt = (20 \text{ m/s} / 4 \text{ s}) = 5 \text{ m/s}^2$

Example 3: Finding Torque
A force of 10 N is applied to a wheel of radius 0.5 m. Find the torque.

$\tau = F \times r = 10 \text{ N} \times 0.5 \text{ m} = 5 \text{ Nm}$

Example 4: Finding Angular Momentum

A wheel of moment of inertia 2 kg m^2 rotates with an angular velocity of 10 rad/s. Find its angular momentum.

$L = I \omega = 2 \text{ kg m}^2 \times 10 \text{ rad/s} = 20 \text{ kg m}^2/\text{s}$

Understanding Rotational Motion:
By understanding these concepts, you can appreciate the beauty and complexity of rotational motion. From the simple spin of a top to the intricate motions of celestial bodies, rotational motion is a fundamental aspect of our physical world.

4. Newton's Laws of Motion

The motion of objects is governed by three fundamental laws established by Sir Isaac Newton, which have revolutionized our understanding of mechanics.

Key Concepts
- **Inertia**: The tendency of an object to resist changes in its motion.
- **External Force**: A force that acts on an object from outside, causing it to change its motion.
- **Constant Velocity**: A velocity that does not change in magnitude or direction

First Law: The Law of Inertia
This law states:
"An object remains at rest or in uniform motion in a straight line unless acted upon by an external force."

Explanation:
Inertia is the property of an object to resist changes in its state of motion. A stationary object does not move unless a force is applied, and a moving object does not stop unless an opposing force acts upon it.

Examples:
a) A book lying on a table remains stationary unless pushed.
b) A football continues rolling on a smooth surface until friction or another force stops it.
c) A Car Moving on a Straight Road: A car moving on a straight road will continue to move with a constant velocity unless acted upon by an external force, such as friction or the brakes.
d) A Ball Thrown Upwards: A ball thrown upwards will continue to move upwards with a decreasing velocity until it reaches its maximum height, at which point it will start to fall downwards due to the external force of gravity.
e) A Satellite in Orbit: A satellite in orbit around the Earth will continue to move with a constant velocity unless acted upon by an external force, such as the gravitational force of the Earth or the thrust of its engines.

Mathematical Operations

The mathematical operation that describes Newton's First Law of Motion is:
$F = 0 \rightarrow a = 0 \rightarrow v = $ constant
where F is the net force acting on an object, a is its acceleration, and v is its velocity.

Applications and Uses

a) **Design of Safety Features**: Newton's First Law of Motion is used in the design of safety features, such as seatbelts and airbags, to protect occupants of vehicles in the event of a crash.

b) **Space Exploration**: Newton's First Law of Motion is used in the design of spacecraft and their trajectories to ensure that they can travel through space with a constant velocity.

c) **Engineering Design**: Newton's First Law of Motion is used in the design of machines and mechanisms to ensure that they can operate with a constant velocity.

Derivations

Here is a diagram that illustrates Newton's First Law of Motion:

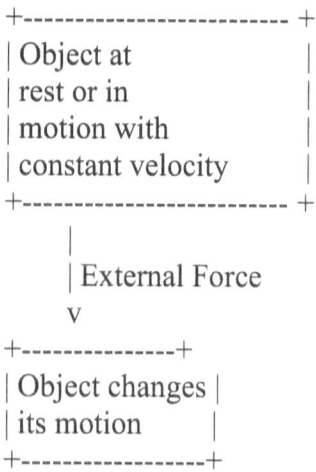

```
+--------------------------+
| Object at                |
| rest or in               |
| motion with              |
| constant velocity        |
+--------------------------+
     |
     | External Force
     v
+-----------------+
| Object changes  |
| its motion      |
+-----------------+
```

Newton's First Law of Motion can be derived

from the concept of inertia and the definition of force.

1. **Inertia**: An object at rest will remain at rest, and an object in motion will continue to move with a constant velocity, unless acted upon by an external force.
2. **Definition of Force**: A force is a push or pull that causes an object to change its motion.

By combining these two concepts, we can derive Newton's First Law of Motion:

$F = 0 \rightarrow a = 0 \rightarrow v = $ constant

Here are some examples from day-to-day life that illustrate Newton's First Law of Motion:
Examples

1. **Seatbelt:** When you're driving a car and suddenly slam on the brakes, you'll continue moving forward until the seatbelt restrains you. This is because your body tends to maintain its state of motion due to inertia.
2. **Coffee Cup**: When you're riding in a car and suddenly turn a corner, your coffee cup will continue moving in a straight line until the force of the turn causes it to change direction.
3. **Bicycle**: When you're riding a bicycle and stop pedaling, you'll continue moving forward until the force of friction slows you down.
4. **Ball on a Billiard Table**: When you hit a ball on a billiard table, it will continue moving in a straight line until it hits another ball or the edge of the table.

5. **Airplane Takeoff**: When an airplane takes off, it must generate enough thrust to overcome its inertia and lift off the ground.
6. **Car Crash**: When two cars collide, the occupants will continue moving forward until the force of the crash causes them to stop.
7. **Gymnast:** When a gymnast is performing a routine, they must use force to change their direction and motion, otherwise, they will continue moving in a straight line.
8. **Skater**: When a skater is gliding on ice, they will continue moving in a straight line until the force of friction slows them down.
9. **Train**: When a train is moving at a constant speed, it will continue moving in a straight line until the force of the brakes causes it to slow down.
10. **Bowling Ball**: When a bowling ball is rolled down a lane, it will continue moving in a straight line until the force of friction slows it down.

Second Law: Force and Acceleration

Newton's second law establishes the relationship between force, mass, and acceleration:

"The acceleration of an object is directly proportional to the net force acting on it and inversely proportional to its mass."

Newton's Second Law of Motion states that the force applied to an object is equal to the mass of the object multiplied by its acceleration. Mathematically, this is expressed as:

$F = ma$

Key Concepts
- **Force (F)**: A push or pull that causes an object to change its motion.
- **Mass (m)**: A measure of the amount of matter in an object.
- **Acceleration (a)**: A change in velocity over time.

Mathematical Expression:

Examples

a. **A Car Accelerating**: When you press the gas pedal in a car, the force of the engine causes the car to accelerate. The more massive the car, the more force is required to produce a given acceleration.
b. **A Ball Thrown**: When you throw a ball, the force of your hand causes the ball to accelerate. The more massive the ball, the more force is required to produce a given acceleration.
c. **A Rocket Launching**: When a rocket launches, the force of the engines causes the rocket to accelerate. The more massive the rocket, the more force is required to produce a given acceleration.

Mathematical Operations

The mathematical operation that describes Newton's Second Law of Motion is:

$F = ma$

This equation can be rearranged to solve for acceleration:

$a = F / m$

Applications and Uses

a. **Design of Vehicles**: Newton's Second Law of Motion is used in the design of vehicles, such as cars, airplanes, and bicycles, to determine the force required to produce a given acceleration.

b. **Rocket Propulsion**: Newton's Second Law of Motion is used in rocket propulsion to determine the force required to produce a given acceleration.

c. **Sports Equipment**: Newton's Second Law of Motion is used in the design of sports equipment, such as golf clubs and tennis rackets, to determine the force required to produce a given acceleration.

Derivations

Newton's Second Law of Motion can be derived from the concept of force and acceleration.

1. **Force**: A force is a push or pull that causes an object to change its motion.

2. **Acceleration**: An acceleration is a change in velocity over time.

By combining these two concepts, we can derive Newton's Second Law of Motion:

$F = ma$

Diagram

Here is a diagram that illustrates Newton's Second

Law of Motion:

```
+---------------+
| Force (F)     |
+---------------+
        |
        | Mass (m)
        v
+---------------+
| Acceleration  |
| (a)           |
+---------------+
```

Vector Operations on Force

1. **Addition of Forces (Vector Addition):**
   Forces are vectors and follow the rules of vector addition. If two forces $\vec{F}_1$ and $\vec{F}_2$ act at a point, the resultant force $\vec{R}$ is the vector sum:

   $$\vec{R} = \vec{F}_1 + \vec{F}_2$$

1. **Subtraction of Forces:**
   Subtraction involves adding the negative of a vector. For $\vec{F}_1 - \vec{F}_2$, this is equivalent to $\vec{F}_1 + (-\vec{F}_2)$, where $-\vec{F}_2$ is a vector with the same magnitude as $\vec{F}_2$ but opposite direction.

3. Multiplication of Forces

- **Scalar Multiplication:** If a force $\vec{F}$ is multiplied by a number k, the result is a vector with magnitude $k \times |\vec{F}|$, pointing in the same direction (if $k > 0$) or the opposite direction (if $k < 0$).

- **Dot Product (Scalar Product):** Measures how much one force projects onto another:

$$\vec{F_1} \cdot \vec{F_2} = |\vec{F_1}||\vec{F_2}| \cos\theta$$

- **Cross Product (Vector Product):** Gives a vector perpendicular to both:

$$\vec{F_1} \times \vec{F_2} = |\vec{F_1}||\vec{F_2}| \sin\theta \, \hat{n}$$

4. Division of Forces

Division isn't directly defined for vectors, but we use ratios or components of vectors to understand relationships between them.

Law of Parallelogram of Forces

The law of parallelogram of forces states that if two forces acting at a point are represented by two adjacent sides of a parallelogram, the resultant is represented in magnitude and direction by the diagonal passing through the same point.

Derivation of the Magnitude and Direction

Consider Two Forces:

- $\vec{F}_1$ and $\vec{F}_2$ act at an angle θ.
- Resultant force: $\vec{R} = \vec{F}_1 + \vec{F}_2$.

Magnitude of Resultant:

Using the vector addition formula and geometry of the parallelogram,

$$|\vec{R}| = \sqrt{|\vec{F}_1|^2 + |\vec{F}_2|^2 + 2|\vec{F}_1||\vec{F}_2|\cos\theta}$$

Direction of Resultant:

Let ϕ be the angle between $\vec{R}$ and $\vec{F}_1$:

$$\tan\phi = \frac{|\vec{F}_2|\sin\theta}{|\vec{F}_1| + |\vec{F}_2|\cos\theta}$$

Final Expressions:

1. **Magnitude of Resultant:**

$$R = \sqrt{F_1^2 + F_2^2 + 2F_1 F_2 \cos\theta}$$

$$\phi = \tan^{-1}\left(\frac{F_2 \sin\theta}{F_1 + F_2 \cos\theta}\right)$$

This approach combines vector principles and geometry to resolve forces effectively.

Understanding Vector Operations on Forces

Forces are vector quantities, meaning they have both magnitude and direction. To understand their interactions, we use the concepts of vector addition, subtraction, and multiplication. Let's break it down step by step.

1. Addition of Forces

When two forces $\vec{F}_1$ and $\vec{F}_2$ act at a point, their combined effect is the **resultant force** $\vec{R}$. Mathematically,

$$\vec{R} = \vec{F}_1 + \vec{F}_2$$

- Place the tail of $\vec{F}_2$ at the head of $\vec{F}_1$.
- Draw a vector from the tail of $\vec{F}_1$ to the head of $\vec{F}_2$. This vector represents $\vec{R}$.

Alternatively, we can use the **parallelogram method**, which we'll explore shortly.

2. Subtraction of Forces

Subtracting vectors involves adding a negative vector. For $\vec{F}_1 - \vec{F}_2$:

$$\vec{F}_1 - \vec{F}_2 = \vec{F}_1 + (-\vec{F}_2)$$

3. Multiplication of Forces

- **Scalar Multiplication:** If a force $\vec{F}$ is multiplied by a number k, the result is a vector with magnitude $k \times |\vec{F}|$, pointing in the same direction (if $k > 0$) or the opposite direction (if $k < 0$).

- **Dot Product (Scalar Product):** Measures how much one force projects onto another:

$$\vec{F}_1 \cdot \vec{F}_2 = |\vec{F}_1||\vec{F}_2|\cos\theta$$

- **Cross Product (Vector Product):** Gives a vector perpendicular to both:

$$\vec{F}_1 \times \vec{F}_2 = |\vec{F}_1||\vec{F}_2|\sin\theta\,\hat{n}$$

4. Division of Forces

Division isn't directly defined for vectors, but we use ratios or components of vectors to understand relationships between them.

The Mechanics of Motion: Force, Friction
and Energy Explored   By: Prashant Kumar Lal

Law of Parallelogram of Forces
This law provides a systematic way to calculate the resultant of two forces acting simultaneously at a point.

Statement of the Law
If two forces acting at a point are represented in magnitude and direction by two adjacent sides of a parallelogram, the diagonal of the parallelogram passing through the point of intersection represents the resultant force in magnitude and direction. $Q = F_1$ and $P = F_2$

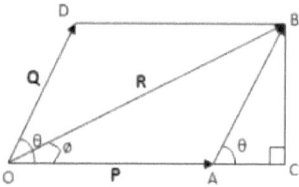

Derivation of the Magnitude and Direction of

1. Let two forces $\vec{F_1}$ and $\vec{F_2}$ act at a point, making an angle θ with each other.

2. Using the **law of cosines** in the parallelogram, the magnitude of the resultant R is:

$$R = \sqrt{F_1^2 + F_2^2 + 2F_1F_2\cos\theta}$$

$$\tan\phi = \frac{F_2\sin\theta}{F_1+F_2\cos\theta}$$

Step-by-Step Explanation

- **Magnitude:**
  Imagine $\vec{F}_1$ and $\vec{F}_2$ forming two sides of a parallelogram. The diagonal is the resultant $\vec{R}$. Use the Pythagoras-like formula to calculate R.

- **Direction:**
  The angle ϕ helps locate the exact orientation of $\vec{R}$. The formula involves trigonometry to account for the angle θ between $\vec{F}_1$ and $\vec{F}_2$.

Finding the Resultant Force

Case 1: Two Forces Acting at 90°

If two forces F_1 and F_2 act at a right angle to each other ($\theta = 90°$), the magnitude of the resultant force R is given by:

$$R = \sqrt{F_1^2 + F_2^2}$$

Case 2: Two Forces Acting at 30°

If two forces F_1 and F_2 act at an angle of $\theta = 30°$, the magnitude of the resultant force R is given by the law of parallelogram of forces:

$$R = \sqrt{F_1^2 + F_2^2 + 2F_1 F_2 \cos 30°}$$

$$R = \sqrt{F_1^2 + F_2^2 + F_1 F_2 \sqrt{3}}$$

Case 3: Many Forces Acting Simultaneously

When multiple forces $\vec{F_1}, \vec{F_2}, \vec{F_3}, \ldots$ act on a point, the resultant is the **vector sum** of all these forces:

$$\vec{R} = \vec{F_1} + \vec{F_2} + \vec{F_3} + \ldots$$

1. Resolve each force into its components along the x- and y-axes:

$$F_{x_{\text{total}}} = F_{1x} + F_{2x} + F_{3x} + \ldots$$

$$R = \sqrt{F^2_{x_{\text{total}}} + F^2_{y_{\text{total}}}}$$

$$\tan \phi = \frac{F_{y_{\text{total}}}}{F_{x_{\text{total}}}}$$

This concept is essential in Physics as it explains how forces combine in nature, such as in balancing a beam or studying the motion of objects under multiple forces.

Law of Polygon of Forces

Statement

The law of polygon of forces states:

If a number of forces acting simultaneously on a

particle are represented in magnitude and direction by the sides of a polygon taken in order, the closing side of the polygon, taken in the opposite order, represents the resultant force in magnitude and direction.

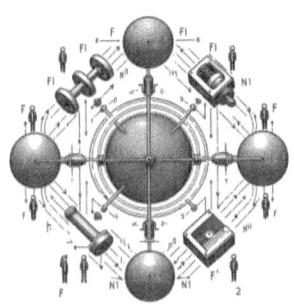

Vector Method

1. Let $\vec{F}_1, \vec{F}_2, \vec{F}_3, \ldots \vec{F}_n$ be the forces acting on a particle.

2. Draw vectors $\vec{F}_1, \vec{F}_2, \ldots$ one after another in sequence such that the tail of each vector starts at the head of the previous one.

3. The resultant force $\vec{R}$ is given by the vector connecting the tail of $\vec{F}_1$ to the head of $\vec{F}_n$, in the opposite order.

The Mechanics of Motion: Force, Friction
and Energy Explored   By: Prashant Kumar Lal

Example Problems

1. **Two Forces at 90°**:
   Forces $F_1 = 3, N$ and $F_2 = 4, N$:

$$R = \sqrt{3^2 + 4^2} = \sqrt{9 + 16} = 5\,N$$

1. **Two Forces at 30°**:
   Forces $F_1 = 5, N$ and $F_2 = 7, N$:

$$R = \sqrt{5^2 + 7^2 + 2(5)(7)\cos 30°}$$

1. **Many Forces**:
   Forces $F_1 = 10, N$ at $0°$, $F_2 = 5, N$ at $90°$, $F_3 = 8, N$ at $180°$:

   ○ Resolve components:

$$F_x = 10 - 8 = 2\,N, \quad F_y = 5$$

$$R = \sqrt{2^2 + 5^2} = \sqrt{4 + 25} = 5.39\,N$$

$$\tan\phi = \tfrac{5}{2}, \quad \phi = \tan^{-1}(2.5) \approx 68.2°$$

To Find Resultant Force using triangle Law of Vectors:

The triangle law of vectors, also known as the triangle rule, is a method used to find the resultant force when two forces act on an object. Here's how to find the resultant force using the triangle law of vectors:

Triangle Law of Vectors:
When two forces, F_1 and F_2, act on an object, the resultant force (R) can be found by constructing a triangle with the two forces as the sides. The resultant force is the third side of the triangle, opposite the angle between the two forces.

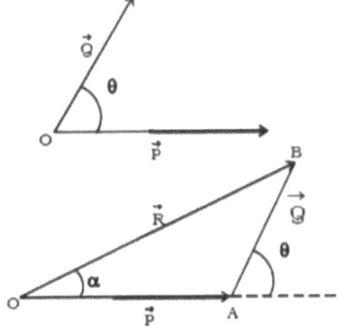

Mathematical Expression:
To deduce the mathematical expression for the resultant force, we can use the law of cosines:
$$R^2 = F_1^2 + F_2^2 + 2F_1F_2 \cos \theta$$
where:
R = resultant force
F_1 = magnitude of force 1
F_2 = magnitude of force 2
θ = angle between the two forces

Solving for Resultant Force:
To find the resultant force, we can take the square root of both sides:
$$R = \sqrt{(F_1^2 + F_2^2 + 2F_1F_2 \cos \theta)}$$

Direction of Resultant Force:
To find the direction of the resultant force, we can use the law of sines:
$$\sin \alpha / F_1 = \sin \beta / F_2 = \sin \gamma / R$$
where:
α = angle between F1 and R
β = angle between F2 and R
γ = angle between F1 and F2

By solving these equations, we can find the direction of the resultant force.

Example:
Suppose we have two forces:
F_1 = 10 N, acting at an angle of 30° with the x-axis
F_2 = 15 N, acting at an angle of 60° with the x-axis

Using the triangle law of vectors, we can find the resultant force:

$R = \sqrt{(10^2 + 15^2 + 2(10)(15)\cos 30°)}$
$= \sqrt{(100 + 225 + 300 \cos 30°)}$
$= \sqrt{(325 + 259.8)}$
$= \sqrt{584.8}$
≈ 24.2 N

The direction of the resultant force can be found using the law of sines:
$\sin \alpha / 10 = \sin \beta / 15 = \sin 30° / 24.2$
Solving these equations, we get:
$\alpha \approx 34.6°$
$\beta \approx 55.4°$

Therefore, the resultant force is approximately 24.2 N, acting at an angle of approximately 34.6° with the x-axis.

Mass vs. Weight

Mass is a measure of the amount of matter in an object, and it is typically denoted by the symbol "m". Mass is an intrinsic property of an object, meaning that it does not change regardless of the object's location or the forces acting upon it.

Weight, on the other hand, is a measure of the force exerted on an object by gravity. Weight is typically denoted by the symbol "W" or "Fg". Weight is a force that depends on both the mass of the object and the acceleration due to gravity (g) at the object's location.

CGS and MKS Units

The CGS (Centimeter-Gram-Second) system and the MKS (Meter-Kilogram-Second) system are two different systems of units used to measure physical quantities.

Here's a comparison of the CGS and MKS units for mass, weight, and force:

Sl. No	Quantity	CGS Unit	MKS Unit
1	Mass	gram (g)	Kilogram (kg)
2	Weight	gram-weight (gw) or dyne	kilogram-weight (kgw) or newton (N)
3	Force	dyne (dyn)	newton (N)

Note that the CGS system is not as commonly used as the MKS system, but it is still used in some contexts.

gm wt and kg wt

The terms "gm wt" and "kg wt" are often used to refer to the weight of an object in grams or kilograms, respectively. However, this notation can be confusing, as it implies that the unit of weight is the same as the unit of mass.

In reality, the unit of weight is typically measured in units of force, such as newtons (N) or dyne (dyn). Therefore, it is more accurate to use the notation "N" or "dyn" to refer to the weight of an object, rather than "gm wt" or "kg wt".

Conversion Factors

Here are some conversion factors that can be used to

convert between different units of mass, weight, and force:
- 1 kg = 1000 g
- 1 N = 100,000 dyn
- 1 kg wt = 9.8 N (approximately)
- 1 gm wt = 9.8 dyn (approximately)

Note that these conversion factors are approximate, as the value of g (the acceleration due to gravity) can vary slightly depending on the location.

Examples:
a) A heavier object requires more force to achieve the same acceleration as a lighter object.
b) Pushing a car with a small force results in slow acceleration, but applying a larger force speeds it up.

c) **A Car Accelerating: A car is accelerating from rest to a speed of 60 km/h in 10 seconds. What is the force required to produce this acceleration if the mass of the car is 1500 kg?**
To solve this problem, we can use Newton's Second Law of Motion, which states that the force (F) applied to an object is equal to the mass (m) of the object multiplied by its acceleration (a).

First, we need to convert the speed from km/h to m/s:
60 km/h = 60,000 m / 3600 s = 16.67 m/s
Next, we can calculate the acceleration (a) of the car:
$a = \Delta v / \Delta t$
$= (16.67 \text{ m/s} - 0 \text{ m/s}) / 10 \text{ s}$

= 1.667 m/s^2

Now, we can use Newton's Second Law of Motion to calculate the force (F) required to produce this acceleration:

F = ma
= 1500 kg × 1.667 m/s^2
= 2500 N

Therefore, the force required to produce this acceleration is 2500 N.

Conceptual Questions

1. What would happen to the acceleration of the car if its mass were doubled?

2. What would happen to the force required to produce the acceleration if the car were accelerated from rest to a speed of 120 km/h in 10 seconds?

3. How would the force required to produce the acceleration change if the car were accelerated on an incline?

Mathematical Operations

1. If the car were accelerated from rest to a speed of 120 km/h in 10 seconds, what would be the new acceleration?

2. If the car were accelerated from rest to a speed of 60 km/h in 5 seconds, what would be the new force required to produce the acceleration?

3. If the car were accelerated on an incline, how would the force required to produce the acceleration change?

Derivations

1. Derive the equation for the force required to produce a given acceleration for an object of mass m.
2. Derive the equation for the acceleration of an object of mass m when a force F is applied to it.
3. Derive the equation for the work done by a force F in moving an object of mass m through a distance d.
3. A Ball Thrown: A ball is thrown upwards with an initial velocity of 20 m/s. What is the force required to produce this acceleration if the mass of the ball is 0.5 kg?

To solve this problem, we need to determine the acceleration of the ball first. Since the ball is thrown upwards, it will experience a downward acceleration due to gravity, which is 9.8 m/s².

However, since the ball is thrown upwards, its initial velocity will be in the opposite direction to the acceleration due to gravity. Therefore, we need to use the equation of motion to determine the acceleration of the ball:

$v = u + at$

where v is the final velocity (0 m/s, since the ball will eventually come to rest), u is the initial velocity (20 m/s), a is the acceleration, and t is the time.

Rearranging the equation to solve for acceleration, we get:

$$a = (v - u) / t$$

Since the ball will eventually come to rest, the final velocity v will be 0 m/s. We can assume that the ball takes about 2 seconds to reach its maximum height and come to rest (this is a rough estimate, but it's good enough for our purposes).

Plugging in the values, we get:
$a = (0 - 20) / 2$
$= -10 \text{ m/s}^2$

The negative sign indicates that the acceleration is in the opposite direction to the initial velocity.
Now that we have the acceleration, we can use Newton's Second Law of Motion to determine the force required to produce this acceleration:
$$F = ma$$
$$= 0.5 \text{ kg} \times -10 \text{ m/s}^2$$
$$= -5 \text{ N}$$
The negative sign indicates that the force is in the opposite direction to the acceleration.

Therefore, the force required to produce this acceleration is 5 N, acting in the downward direction.

Conceptual Questions

1. What would happen to the acceleration of the ball if its mass were doubled?
2. What would happen to the force required to produce the acceleration if the ball were thrown upwards with an initial velocity of 40 m/s?
3. How would the force required to produce the

acceleration change if the ball were thrown on a planet with a different gravitational acceleration?

Mathematical Operations

1. If the ball were thrown upwards with an initial velocity of 40 m/s, what would be the new acceleration?
2. If the ball were thrown on a planet with a gravitational acceleration of 5 m/s^2, what would be the new force required to produce the acceleration?
3. If the ball were thrown with an initial velocity of 20 m/s at an angle of 30° above the horizontal, what would be the new acceleration and force required to produce it?

Derivations

1. Derive the equation for the force required to produce a given acceleration for an object of mass m.
2. Derive the equation for the acceleration of an object of mass m when a force F is applied to it.
3. Derive the equation for the work done by a force F in moving an object of mass m through a distance d.
4. A Rocket Launching: A rocket is launching into space with an acceleration of 10 m/s^2. What is the force required to produce this acceleration if the mass of the rocket is 1000 kg?

To solve this problem, we can use Newton's Second Law of Motion, which states that the force (F) applied to an object is equal to the mass (m) of the object multiplied by its acceleration (a).
Given values:

m = 1000 kg (mass of the rocket)
a = 10 m/s² (acceleration of the rocket)

We can plug these values into the equation F = ma:

F = 1000 kg × 10 m/s²
= 10,000 N
Therefore, the force required to produce this acceleration is 10,000 N.

Conceptual Questions
1. What would happen to the acceleration of the rocket if its mass were doubled?
2. What would happen to the force required to produce the acceleration if the rocket were launched with an acceleration of 20 m/s²?
3. How would the force required to produce the acceleration change if the rocket were launched on a planet with a different gravitational acceleration?

Mathematical Operations
1. If the rocket were launched with an acceleration of 20 m/s², what would be the new force required to produce the acceleration?
2. If the rocket were launched on a planet with a gravitational acceleration of 5 m/s², what would be the new force required to produce the acceleration?
3. If the rocket were launched with an initial velocity of 100 m/s and an acceleration of 10 m/s², what would be the new force required to produce the acceleration?

Derivations
1. Derive the equation for the force required to produce

a given acceleration for an object of mass m.
2. Derive the equation for the acceleration of an object of mass m when a force F is applied to it.
3. Derive the equation for the work done by a force F in moving an object of mass m through a distance d.

Diagram
Here is a diagram that illustrates the rocket launching into space:

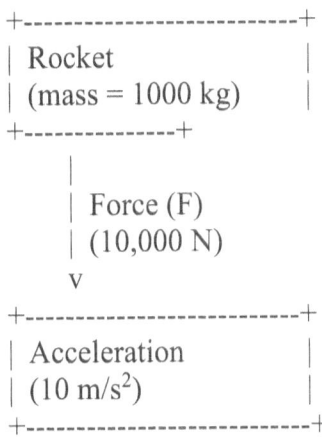

These conceptual problems can help you *understand Newton's Second Law of Motion and its applications in real-life situations.*

Derivation of Newton's Second Law using Calculus, momentum and impulse

Newton's Second Law states that the force (F) applied to an object is equal to the mass (m) of the object multiplied by its acceleration (a). Mathematically, this is expressed as:

$F = ma$

To derive this equation using calculus, we can start with the definition of acceleration:

$a = dv/dt$

where a is the acceleration, v is the velocity, and t is time.

We can then multiply both sides of the equation by the mass (m) of the object:

$ma = m(dv/dt)$

Next, we can use the product rule of differentiation to expand the right-hand side of the equation:

$ma = d(mv)/dt$

Now, we can define the momentum (p) of the object as:

$p = mv$

Substituting this definition into the previous equation, we get:

$ma = dp/dt$

Finally, we can define the force (F) applied to the object as:

$F = dp/dt$

Substituting this definition into the previous equation, we get:

$F = ma$

which is Newton's Second Law.

Momentum

Momentum is a measure of the amount of motion an object has. It is defined as the product of an object's mass and velocity:

$p = mv$

Momentum is a vector quantity, meaning it has both magnitude and direction.

Derivation of Units of Momentum

Momentum (p) is defined as the product of an object's mass (m) and velocity (v):

$p = mv$

The unit of mass is typically measured in kilograms (kg) and the unit of velocity is typically measured in meters per second (m/s).

Therefore, the unit of momentum is:

$p = kg \times m/s = kg \cdot m/s$

This unit is often referred to as a kilogram-meter per second (kg·m/s).

Mathematical Proof that the Rate of Change of Velocity is Force

We can start with the definition of acceleration (a) as the rate of change of velocity (v) with respect to time (t):

$a = dv/dt$

We can then multiply both sides of the equation by the mass (m) of the object:

$ma = m(dv/dt)$

The left-hand side of the equation is the force (F) applied to the object, according to Newton's Second Law:

$F = ma$

Therefore, we can rewrite the equation as:

$F = m(dv/dt)$

This equation shows that the force applied to an object is equal to the mass of the object multiplied by the rate of change of its velocity.

Alternative Derivation using the Concept of Momentum

We can also derive the result using the concept of momentum. We can start with the definition of momentum (p) as the product of an object's mass (m) and velocity (v):
$p = mv$
We can then take the derivative of both sides of the equation with respect to time (t):
$dp/dt = m(dv/dt)$
The left-hand side of the equation is the rate of change of momentum, which is equal to the force (F) applied to the object:
$F = dp/dt$
Therefore, we can rewrite the equation as:
$F = m(dv/dt)$
This equation shows that the force applied to an object is equal to the mass of the object multiplied by the rate of change of its velocity.

Impulse
Impulse is a measure of the change in momentum of an object over a given time period. It is defined as the product of the average force applied to an object and the time over which the force is applied:
$J = F\Delta t$
Impulse is also a vector quantity, meaning it has both magnitude and direction.
Units of Impulse

The unit of force is typically measured in newtons (N) and the unit of time is typically measured in seconds (s).
Therefore, the unit of impulse is:
J = N × s = N·s
This unit is often referred to as a newton-second (N·s).

Alternative Derivation using the Concept of Momentum

We can also derive the unit of impulse using the concept of momentum. We can start with the definition of impulse as the change in momentum:
$J = \Delta p$
We can then substitute the definition of momentum:
$J = \Delta(mv)$
Since the mass (m) is constant, we can pull it out of the derivative:
$J = m\Delta v$
The unit of mass is typically measured in kilograms (kg) and the unit of velocity is typically measured in meters per second (m/s).
Therefore, the unit of impulse is:
J = kg × m/s = kg·m/s
Since 1 N = 1 kg·m/s², we can rewrite the unit of impulse as:
J = N·s

Applications and Uses

1. Rocket Propulsion: Momentum and impulse are crucial in rocket propulsion, where the goal is to generate a high velocity and momentum to escape Earth's gravity.
2. Car Safety: Momentum and impulse are important in

car safety, where the goal is to minimize the force of impact and reduce the change in momentum.

3. Sports: Momentum and impulse are essential in many sports, such as football, hockey, and tennis, where the goal is to generate a high velocity and momentum to score points or win games.

Day-to-Day Examples

1. A Car Crash: When a car crashes into a wall, the momentum of the car is suddenly reduced to zero, resulting in a large impulse and a significant force of impact.

2. A Baseball Pitch: When a baseball pitcher throws a ball, the momentum of the ball is generated by the force of the pitcher's arm and the time over which the force is applied.

3. A Rocket Launch: When a rocket is launched into space, the momentum of the rocket is generated by the force of the engines and the time over which the force is applied.

Conceptual Problems

1. A Car Traveling at Constant Velocity: If a car is traveling at a constant velocity, what is the net force acting on the car?

2. A Ball Thrown Upwards: If a ball is thrown upwards with an initial velocity, what is the momentum of the ball at its highest point?

3. A Rocket Accelerating in Space: If a rocket is accelerating in space with a constant force, what is the momentum of the rocket after a given time period?

Mathematical Approach

1. Momentum and Velocity: Derive the equation for momentum in terms of velocity and mass.
2. Impulse and Force: Derive the equation for impulse in terms of force and time.
3. Momentum and Impulse: Derive the equation for the change in momentum of an object in terms of the impulse applied to the object.

Mathematical Operations
1. Momentum and Velocity: If the velocity of an object is doubled, what happens to its momentum?
2. Impulse and Force: If the force applied to an object is tripled, what happens to the impulse?
3. Momentum and Impulse: If the impulse applied to an object is doubled, what happens to its momentum?

Third Law: Action and Reaction

Newton's third law states:
"For every action, there is an equal and opposite reaction."
Explanation:
Whenever one object exerts a force on another, the second object exerts an equal force in the opposite direction.
Examples:
   a) When a rocket launches, the exhaust gases push downward while the rocket moves upward.
   b) A Car Accelerating: When a car accelerates forward, the wheels of the car push backward on the road, and the road pushes forward on

the wheels. This is an example of Newton's Third Law in action.

c) **A Ball Bouncing:** When a ball bounces off the ground, the ball exerts a force downward on the ground, and the ground exerts an equal and opposite force upward on the ball.
d) **A Swimmer Swimming:** When a swimmer swims through the water, the swimmer exerts a force backward on the water, and the water exerts an equal and opposite force forward on the swimmer.

Mathematical Operations
Newton's Third Law can be mathematically represented as:
$F_1 = -F_2$
where F_1 is the force exerted by object A on object B, and F_2 is the force exerted by object B on object A.

Applications and Uses
1. Rocket Propulsion: Newton's Third Law is used in rocket propulsion, where the hot gases expelled from the back of the rocket exert a forward force on the rocket.
2. Jet Engines: Newton's Third Law is used in jet engines, where the hot gases expelled from the back of the engine exert a forward force on the plane.
3. Hydraulic Systems: Newton's Third Law is used in hydraulic systems, where the pressure exerted by a fluid on a piston is equal to the pressure exerted by the piston on the fluid.

Derivations

Newton's Third Law can be derived from the concept of momentum and the law of conservation of momentum.

Force: The Cause of Motion

Force is any interaction that can change the motion of an object. It is a vector quantity, meaning it has both magnitude and direction.

Diagram

Here is a diagram that illustrates Newton's Third Law:

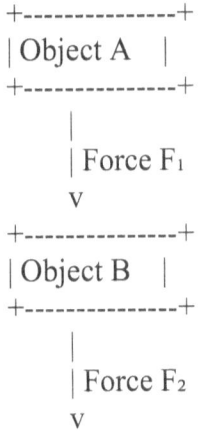

A mathematical derivation of Newton's Third Law from the concept of the Law of Conservation of Momentum:

Law of Conservation of Momentum

The Law of Conservation of Momentum states that the total momentum of a closed system remains constant over time. Mathematically, this can be expressed as:

$\Delta p = 0$
where Δp is the change in momentum.

Mathematical Proof of the Law of Conservation of Momentum

To prove the Law of Conservation of Momentum mathematically, we can consider a closed system consisting of two objects, A and B, that interact with each other.

Let's assume that object A has a mass of m_1 and an initial velocity of v_1, and object B has a mass of m_2 and an initial velocity of v_2.
The initial momentum of object A is:
$p_1 = m_1 v_1$
The initial momentum of object B is:
$p_2 = m_2 v_2$
The total initial momentum of the system is:
$p_0 = p_1 + p_2 = m_1 v_1 + m_2 v_2$
After the interaction, the final momentum of object A is:
$p_1' = m_1 v_1'$
The final momentum of object B is:
$p_2' = m_2 v_2'$
The total final momentum of the system is:
$p' = p_1' + p_2' = m_1 v_1' + m_2 v_2'$
Since the system is closed, the total momentum remains constant:
$p_0 = p'$
Substituting the expressions for p_0 and p', we get:
$m_1 v_1 + m_2 v_2 = m_1 v_1' + m_2 v_2'$

Rearranging the equation, we get:
$$m_1(v_1' - v_1) = -m_2(v_2' - v_2)$$
Dividing both sides by Δt, we get:
$$m_1 a_1 = -m_2 a_2$$
where a_1 and a_2 are the accelerations of objects A and B, respectively.

Force and Acceleration

Using Newton's Second Law, we can express the force acting on object A as:
$$F_1 = m_1 a_1$$
Similarly, the force acting on object B is:
$$F_2 = m_2 a_2$$
Substituting these expressions into the previous equation, we get:
$$F_1 = -F_2$$
This equation shows that the force exerted by object A on object B is equal in magnitude and opposite in direction to the force exerted by object B on object A.

Newton's Third Law

This equation is a mathematical statement of Newton's Third Law:
"For every action, there is an equal and opposite reaction."
In this case, the action is the force exerted by object A on object B, and the reaction is the force exerted by object B on object A.

Conceptual Problems

1. A Car Crash: When a car crashes into a wall, what is the reaction force exerted by the wall on the car?
2. A Ball Thrown: When a ball is thrown upwards,

what is the reaction force exerted by the air on the ball?

3. A Swimmer Swimming: When a swimmer swims through the water, what is the reaction force exerted by the water on the swimmer?

Mathematical Operations

1. Force and Acceleration: If a car accelerates from rest to a speed of 60 km/h in 10 seconds, what is the reaction force exerted by the road on the car?

2. Momentum and Velocity: If a ball has a mass of 0.5 kg and a velocity of 20 m/s, what is the reaction force exerted by the air on the ball?

3. Energy and Work: If a swimmer swims a distance of 100 meters in 2 minutes, what is the reaction force exerted by the water on the swimmer?

Rocket propulsion, including the principle, working, and mathematical explanation

Principle of Rocket Propulsion

Rocket propulsion is based on the principle of

conservation of momentum, which states that the total momentum of a closed system remains constant over time.

Working of Rocket Propulsion

A rocket works by expelling hot gases out of the back of the rocket, which produces a forward force that propels the rocket upward. The hot gases are produced by burning fuel, such as liquid hydrogen or kerosene, in a combustion chamber.

Here's a step-by-step explanation of the working of a rocket:

1. **Fuel and Oxidizer**: The rocket's fuel tank contains a fuel, such as liquid hydrogen, and an oxidizer, such as liquid oxygen.
2. **Combustion Chamber**: The fuel and oxidizer are pumped into a combustion chamber, where they are burned to produce hot gases.
3. **Nozzle**: The hot gases are expelled out of the back of the rocket through a nozzle, which accelerates the gases to high speed.
4. **Thrust**: As the hot gases are expelled out of the back of the rocket, they produce a forward force that propels the rocket upward. This force is called thrust.
5. **Rocket Acceleration**: As the rocket expels hot gases out of the back, it accelerates upward, gaining speed and altitude.

Mathematical Explanation

The thrust produced by a rocket can be calculated using the following equation:

$F = (m_0 \times v_e) + (p_e - p_0) \times A_e$

Where:

- F is the thrust produced by the rocket
- m_0 is the mass flow rate of the exhaust gases
- v_e is the exhaust velocity of the gases
- p_e is the pressure of the exhaust gases
- p_0 is the ambient pressure
- A_e is the area of the nozzle exit

The exhaust velocity of the gases can be calculated using the following equation:

$$V_e = \sqrt{((2 \times \gamma \times R \times T_e) / (M \times (\gamma - 1)))}$$

Where:
- v_e is the exhaust velocity of the gases
- γ is the adiabatic index of the gas
- R is the molar gas constant
- T_e is the temperature of the exhaust gases
- M is the molecular weight of the gas

The law of conservation of mass and mass defect, along with a mathematical calculation of mass defect:

Law of Conservation of Mass

The law of conservation of mass states that matter cannot be created or destroyed in a chemical reaction. This means that the total mass of the reactants is equal to the total mass of the products.

Mass Defect

Mass defect is the difference between the mass of a nucleus and the sum of the masses of its individual nucleons (protons and neutrons). This difference arises because the binding energy that holds the nucleus together is released as energy, and this energy is equivalent to a certain amount of mass, according to

Einstein's famous equation $E=mc^2$.

Mathematical Calculation of Mass Defect

Let's consider an example of the nucleus of helium-4 (^{4}He), which consists of two protons and two neutrons. The mass of a proton is approximately 1.007276 atomic mass units (amu), and the mass of a neutron is approximately 1.008665 amu.
The total mass of the individual nucleons is:
2(1.007276 amu) + 2(1.008665 amu) = 4.031882 amu
However, the actual mass of the ^{4}He nucleus is approximately 4.002603 amu.
The mass defect is the difference between the two masses:
Mass defect = 4.031882 amu - 4.002603 amu = 0.029279 amu
This mass defect is equivalent to a certain amount of energy, which is released as binding energy that holds the nucleus together.
Using Einstein's equation $E=mc^2$, we can calculate the energy equivalent of the mass defect:
Energy = (0.029279 amu) x (1.66053904 × 10^{-27} kg/amu) x (2.99792458 × 10^8 m/s)^2
Energy ≈ 2.722 × 10^{-12} J
This energy is the binding energy that holds the ^{4}He nucleus together.

Derivation of a mathematical expression for mass defect:

Let's consider a nucleus consisting of Z protons and N neutrons. The total mass of the individual nucleons is:

$m_{total} = Zm_p + Nm_n$

where m_p is the mass of a proton and m_n is the mass of a neutron.

The actual mass of the nucleus is:
$m_{nucleus}$

The mass defect is the difference between the two masses:

$\Delta m = m_{total} - m_{nucleus}$

Substituting the expressions for m_{total} and $m_{nucleus}$, we get:

$\Delta m = Zm_p + Nm_n - m_{nucleus}$

We can express the mass defect in terms of the binding energy (E_b) that holds the nucleus together:

$\Delta m = E_b / c^2$

where c is the speed of light.

Substituting this expression into the previous equation, we get:

$E_b / c^2 = Zm_p + Nm_n - m_{nucleus}$

Multiplying both sides by c^2, we get:

$E_b = (Zm_p + Nm_n - m_{nucleus}) \times c^2$

This is the mathematical expression for the binding energy of a nucleus in terms of the mass defect.
Rearranging the equation to solve for the mass defect, we get:

$\Delta m = (E_b / c^2)$

This is the mathematical expression for the mass defect of a nucleus in terms of its binding energy.

Chapter - 2

Force and Its Many Forms

Types of Forces

Forces are pushes or pulls that can cause an object to change its motion or shape. There are several types of forces, which can be classified based on their strength, range, and origin.

Fundamental Forces of Nature
There are four fundamental forces of nature:
a) Gravitational Force: a long-range force that attracts two objects with mass towards each other.
b) Electromagnetic Force: a long-range force that acts between charged particles, such as protons and electrons.
c) Strong Nuclear Force: a short-range force that holds quarks together inside protons and neutrons, and holds these particles together

inside atomic nuclei.
d) Weak Nuclear Force: a short-range force that is responsible for certain types of radioactive decay, where a nucleus emits particles to become more stable.

Classification of Forces

Forces can also be classified into different categories based on their origin and characteristics:

Natural Forces

Natural forces are forces that occur naturally in the environment, without any human intervention. Examples include:
- Gravity
- Friction
- Normal force
- Tension
- Air resistance

Unnatural Forces

Unnatural forces, on the other hand, are forces that are created by human beings, such as:
- Applied force (e.g., pushing or pulling an object)
- Spring force
- Magnetic force
- Electric force

Weak Forces

Weak forces are forces that are relatively weak compared to other forces. Examples include:
- Weak nuclear force
- Magnetic force
- Electric force (in certain situations)

Strong Forces

Strong forces, on the other hand, are forces that are relatively strong compared to other forces. Examples include:
- Strong nuclear force
- Gravitational force (in certain situations)
- Friction (in certain situations)

Note that the classification of forces into weak and strong forces is relative, and can depend on the specific situation or context.

Impact of Forces

Forces have a significant impact on our daily lives and the world around us. They are responsible for:

> a. Motion: Forces cause objects to move, change direction, or change speed.
> b. Shape and Structure: Forces hold objects together, giving them shape and structure.
> c. Energy Transfer: Forces transfer energy from one object to another.
> d. Change in Momentum: Forces cause a change in momentum, which is the product of an object's mass and velocity.

Necessities of Forces

Forces are necessary for:
a. Life: Forces are essential for life, as they enable us to move, breathe, and maintain our bodily functions.
b. Technology: Forces are crucial for technological

advancements, such as transportation, communication, and energy generation.
c. Ecosystems: Forces play a vital role in maintaining ecosystems, such as the water cycle, climate patterns, and the movement of tectonic plates.

Gravitational forces, frictional forces, normal forces, and tension forces, along with examples, mathematical operations, applications, uses, derivations, diagrams, and conceptual problems:

Gravitational Forces

Gravitational forces are attractive forces that exist between two objects with mass. The force of gravity is proportional to the product of the masses of the two objects and inversely proportional to the square of the distance between them.

Mathematical Operation:

$F_g = (G \times m_1 \times m_2) / r^2$

where F_g is the gravitational force, G is the gravitational constant (6.67408^{-11} N m² kg^{-2}), m_1 and m_2 are the masses of the two objects, and r is the distance between them.

Examples:

1. The force of gravity between the Earth and an object on its surface.

2. The force of gravity between the Earth and the Moon.

Applications:

1. Understanding the motion of planets and stars in our

solar system.
2. Calculating the weight of an object on different planets.

Derivation:
The gravitational force can be derived from Newton's law of universal gravitation, which states that every point mass attracts every other point mass by a force acting along the line intersecting both points.

Diagram:

A diagram showing the gravitational force between two objects with mass:

Object 1 (m_1)---------------Object 2 (m_2)

F_g

Conceptual Problems:
1. Why do objects on the surface of the Earth experience a downward force?
2. How does the force of gravity between two objects change if the distance between them is doubled?

Frictional Forces

Frictional forces are forces that oppose the motion of an object. They arise from the interaction between two surfaces that are in contact.
Mathematical Operation:

$F_f = \mu \times N$

where F_f is the frictional force, μ is the coefficient of friction, and N is the normal force.

Examples:
1. The force of friction between a car's tires and the road.
2. The force of friction between a block of wood and a rough surface.

Applications:
1. Understanding the motion of objects on different surfaces.
2. Designing systems to reduce friction, such as lubricants and bearings.

Derivation:
The frictional force can be derived from the concept of surface roughness and the interaction between two surfaces.

Diagram:
A diagram showing the frictional force between two surfaces:

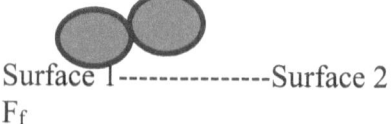

Surface 1---------------Surface 2
F_f

Conceptual Problems:
1. Why do objects experience a force opposing their motion when they are in contact with a surface?
2. How does the force of friction change if the surface roughness is increased?

Normal Forces

Normal forces are forces that act perpendicular to a surface. They arise from the interaction between an object and a surface.

Mathematical Operation:
N = m x g
where N is the normal force, m is the mass of the object, and g is the acceleration due to gravity.

When an object is placed on a surface, it experiences a normal force exerted by the surface. The normal force is perpendicular to the surface and can be resolved into two components: one parallel to the surface and one perpendicular to the surface.

Let's consider an object placed on a surface at an angle θ to the horizontal. The normal force N can be resolved into two components:
N_x = N x sin(θ) (parallel to the surface)
N_y = N x cos(θ) (perpendicular to the surface)
The component N_x is the horizontal component of the normal force, while the component N_y is the vertical component of the normal force.

Example Problems
1. An object is placed on a surface at an angle of 45° to the horizontal. If the normal force exerted by the surface is 50 N, what are the horizontal and vertical components of the normal force?
Solution: N_x = 50 x sin(45°) = 35.35 N
N_y = 50 x cos(45°) = 35.35 N

Examples:
1. The force exerted by the ground on an object placed on it.

2. The force exerted by a table on an object placed on it.

Applications:
1. Understanding the motion of objects on different surfaces.
2. Designing systems to support heavy loads.

Derivation:
The normal force can be derived from the concept of surface interaction and the weight of an object.
Diagram:
A diagram showing the normal force between an object and a surface:

Object---------------Surface

N

Conceptual Problems:
1. Why do objects experience a force perpendicular to a surface when they are in contact with it?
2. How does the normal force change if the mass of the object is increased?

Tension Forces

Tension forces are forces that act along a string or cable. They arise from the interaction between the string or cable and the objects attached to it.

Mathematical Operation:
$T = m \times g$
where T is the tension force, m is the mass of the object,

and g is the acceleration due to gravity.

When a string or cable is attached to an object, it exerts a tension force on the object. The tension force can be resolved into two components: one parallel to the string and one perpendicular to the string.

Let's consider a string attached to an object at an angle θ to the horizontal. The tension force T can be resolved into two components:

$T_x = T \times \cos(\theta)$ (parallel to the string)

$T_y = T \times \sin(\theta)$ (perpendicular to the string)

The component T_x is the horizontal component of the tension force, while the component T_y is the vertical component of the tension force.

1. A string is attached to an object at an angle of 30° to the horizontal. If the tension force in the string is 20 N, what are the horizontal and vertical components of the tension force?

Solution: $T_x = 20 \times \cos(30°) = 17.32$ N

$T_y = 20 \times \sin(30°) = 10$ N

Examples:

1. The force exerted by a string on an object attached to it.
2. The force exerted by a cable on a lift.

Applications:

1. Understanding the motion of objects attached to strings or cables.
2. Designing systems to lift heavy loads.

Derivation:

The tension force can be derived from the concept of surface interaction and the weight of an object.

Diagram:

A diagram showing the tension force along a string:

Object---------------String
T
Conceptual Problems:

1. Why do objects experience a force along a string or cable when they are attached to it?
2. How does the tension force change if the mass of the object is increased?

Note: These explanations are detailed and include

Solved Numerical:
1. Give the magnitude and direction of the net force acting on a stone of mass 0.1 kg which is tied to a string of length 0.25 m and is rotated in a horizontal circle with a speed of 2 m/s.
Answer: The net force acting on the stone is directed towards the center of the circle. The magnitude of the net force is given by:
$F = (m \times v^2) / r$
$= (0.1 \text{ kg} \times (2 \text{ m/s})^2) / 0.25 \text{ m}$
$= 1.6 \text{ N}$

2. A constant retarding force of 50 N is applied to a body of mass 20 kg moving with an initial speed of 15 m/s. How long does the body take to stop?
Answer: Using Newton's second law of motion, we can write:
$F = m \times a$
$50 \text{ N} = 20 \text{ kg} \times a$
$a = -2.5 \text{ m/s}^2$ (retarding force)

Using the equation of motion:
$v = u + at$
$0 = 15 \text{ m/s} + (-2.5 \text{ m/s}^2) \times t$
$t = 6 \text{ s}$

3. A body of mass 5 kg is acted upon by two forces of magnitude 8 N and 6 N, inclined at an angle of 60° with each other. Calculate the resulting force and its direction.

Answer: Using the parallelogram law of forces, we can write:
$R = \sqrt{(F_1^2 + F_2^2 + 2F_1F_2\cos\theta)}$

$R = \sqrt{(8^2 + 6^2 + 2 \times 8 \times 6 \times \cos 60°)}$
$= \sqrt{(64 + 36 + 48)}$
$= \sqrt{148}$
$= 12.16 \text{ N}$

The direction of the resulting force can be found using the equation:
$\tan\alpha = (F_2\sin\theta) / (F_1 + F_2\cos\theta)$
$\tan\alpha = (6 \times \sin 60°) / (8 + 6 \times \cos 60°)$
$= 0.577$
$\alpha = 30°$

4. An object of mass 2 kg is moving with an initial velocity of 10 m/s. A force of 6 N acts on the object for 2 s. Calculate the change in momentum.

Answer: Using the equation:
$\Delta p = F \times \Delta t$
$\Delta p = 6 \text{ N} \times 2 \text{ s}$
$= 12 \text{ kg m/s}$

5. A force of 10 N acts on an object of mass 5 kg for 2 s. Calculate the impulse and the change in momentum.

Answer: Using the equation:

$J = F \times \Delta t$

$J = 10 \text{ N} \times 2 \text{ s}$
$= 20 \text{ N s}$

The change in momentum is equal to the impulse:
$\Delta p = J$
$= 20 \text{ kg m/s}$

6. A force of 10 N acts on an object of mass 2 kg for 3 s. Calculate the change in momentum and the impulse.

Answer: Using the equation:
$\Delta p = F \times \Delta t$
$\Delta p = 10 \text{ N} \times 3 \text{ s}$
$= 30 \text{ kg m/s}$

The impulse is equal to the change in momentum:
$J = \Delta p$
$= 30 \text{ N s}$

7. An object of mass 5 kg is moving with an initial velocity of 2 m/s. A force of 10 N acts on the object for 2 s. Calculate the change in momentum and the impulse.

Answer: Using the equation:
$\Delta p = F \times \Delta t$
$\Delta p = 10 \text{ N} \times 2 \text{ s}$
$= 20 \text{ kg m/s}$

The impulse is equal to the change in momentum:
$J = \Delta p$
$= 20 \text{ N s}$

8. A particle moving in a plane has a velocity vector

given by v = (2t i + 3t² j) m/s, where t is the time in seconds. Find the acceleration vector at t = 2 s.
Answer: Using the equation:
a = dv/dt
a = d (2t i + 3t² j)/dt
= 2 i + 6t j
At t = 2 s:
a = 2 i + 12 j
= $\sqrt{(2^2 + 12^2)}$
= 12.65 m/s²

9. A particle moving in a plane has a position vector given by r = (3t² i + 2t j) m, where t is the time in seconds. Find the velocity and acceleration vectors at t = 2 s.

To find the velocity and acceleration vectors, we need to differentiate the position vector with respect to time.

Velocity Vector

The velocity vector is given by:
v = dr/dt
= d (3t² i + 2t j)/dt
= 6t i + 2 j
At t = 2 s:
v = 6(2) i + 2 j
= 12 i + 2 j

The magnitude of the velocity vector is:
|v| = $\sqrt{(12^2 + 2^2)}$
= $\sqrt{(144 + 4)}$
= $\sqrt{148}$
= 12.16 m/s

Acceleration Vector

The acceleration vector is given by:

$a = dv/dt$

$= d(6t\,i + 2\,j)/dt$

$= 6\,i$

The acceleration vector is constant and directed along the x-axis.

At $t = 2$ s, the acceleration vector remains the same:

$a = 6\,i$

The magnitude of the acceleration vector is:

$|a| = \sqrt{(6^2)}$

$= \sqrt{36}$

$= 6$ m/s²

Here are ten conceptual questions with their answers on the topics of forces, Newton's laws of motion, and motion in a plane:

1. What is the difference between a force and a pressure?

Answer: A force is a push or pull that acts on an object, while pressure is the force per unit area exerted on an object.

2. A car is moving at a constant velocity on a straight road. What is the net force acting on the car?

Answer: The net force acting on the car is zero, as the car is moving at a constant velocity.

3. What is the normal force acting on an object placed on a horizontal surface?

Answer: The normal force acting on an object placed

on a horizontal surface is equal to the weight of the object.

4. A particle is moving in a circular path with a constant speed. What is the direction of the acceleration of the particle?

Answer: The direction of the acceleration of the particle is towards the center of the circle.

5. What is the impulse of a force acting on an object?

Answer: The impulse of a force acting on an object is equal to the change in momentum of the object.

6. A car is accelerating from rest to a speed of 20 m/s in 4 s. What is the acceleration of the car?

Answer: The acceleration of the car is 5 m/s².

7. What is the force acting on an object moving with a constant velocity?

Answer: The force acting on an object moving with a constant velocity is zero.

8. A block of mass 10 kg is placed on a rough horizontal surface. The coefficient of friction between the block and the surface is 0.2. What is the force of friction acting on the block when it is moving with a constant velocity?

Answer: The force of friction acting on the block is 20 N.

9. A force of 10 N acts on an object of mass 2 kg for 3 s. What is the change in momentum of the object?

Answer: The change in momentum of the object is equal to the impulse of the force, which is 30 Ns. Since the mass of the object is 2 kg, the change in velocity is 15 m/s.

10. A car is moving around a circular track of radius 50 m with a constant speed of 20 m/s. What is the centripetal force acting on the car?

Answer: The centripetal force acting on the car is directed towards the center of the circle and is given by the equation Fc = (m × v²) / r. Substituting the values, Fc = (1500 kg × (20 m/s) ²) / 50 m = 12000 N.

Here is an exercise on the topics of forces, Newton's laws of motion, and motion in a plane:

Numerical Questions

1. A force of 20 N acts on an object of mass 5 kg. What is the acceleration of the object? (Hint: Use F = ma)

2. A car is moving with a velocity of 30 m/s. If a force of 100 N is applied to the car for 2 s, what is the change in momentum of the car? (Hint: Use $\Delta p = F \times \Delta t$)

3. An object of mass 10 kg is placed on a rough horizontal surface. The coefficient of friction between the object and the surface is 0.2. What is the force of friction acting on the object when it is moving with a constant velocity? (Hint: Use $F_f = \mu \times N$)

4. A particle is moving in a circular path with a constant speed of 20 m/s. If the radius of the circle is 50 m, what is the centripetal force acting on the particle? (Hint: Use $F_c = (m \times v^2) / r$)

5. A force of 50 N acts on an object of mass 8 kg for

3 s. What is the impulse of the force? (Hint: Use $J = F \times \Delta t$)

6. A car is moving with a velocity of 40 m/s. If a force of 200 N is applied to the car for 1 s, what is the change in velocity of the car? (Hint: Use $\Delta v = (F \times \Delta t) / m$)

7. An object of mass 15 kg is placed on a smooth horizontal surface. What is the normal force acting on the object? (Hint: Use $N = m \times g$)

8. A particle is moving in a circular path with a constant speed of 30 m/s. If the radius of the circle is 75 m, what is the centripetal acceleration of the particle? (Hint: Use $a_c = v^2 / r$)

9. A force of 80 N acts on an object of mass 10 kg for 2 s. What is the change in momentum of the object? (Hint: Use $\Delta p = F \times \Delta t$)

10. A car is moving with a velocity of 50 m/s. If a force of 300 N is applied to the car for 2 s, what is the change in kinetic energy of the car? (Hint: Use $\Delta KE = (F \times \Delta t) \times v$)

Short Questions

1. What is the difference between a force and a pressure?

2. What is the law of inertia?

3. What is the force of friction?

4. What is the centripetal force?

5. What is the difference between a conservative force and a non-conservative force?

6. What is the work-energy theorem?

7. What is the power of a force?

8. What is the torque of a force?

9. What is the angular momentum of a particle?

10. What is the law of universal gravitation?

Very Short Questions

1. What is the unit of force?

2. What is the symbol for acceleration?

3. What is the law of conservation of momentum?

4. What is the centripetal acceleration?

5. What is momentum?

6. What is impulse?

7. What is mass defect?

8. What is the air resistance?

9. What is the buoyant force?

10. What is the gravitational force?

Long Answer Questions

1. Derive the equation of motion for an object under the influence of a constant force.

2. Prove that the law of conservation of momentum is true? State the law of conservation of momentum.

3. Wha is the centripetal force? Derive the equation for centripetal force and explain its significance.

2. Explain the law of conservation of momentum. Provide examples to illustrate the concept.

4. What is the difference between a conservative force and a non-conservative force? Provide examples to illustrate the difference.

5. Prove that $V^2 = u^2 + 2as$. Symbols have their usual meaning

Here are ten questions on the topics of forces, Newton's laws of motion, and motion in a plane, based on the format "What happens when...":

1. What happens when a force is applied to an object at rest?

(Answer: The object accelerates in the direction of the force.)

2. What happens when the net force acting on an object is zero?

(Answer: The object remains at rest or moves with a constant velocity.)

3. What happens when a car turns a corner on a circular path?

(Answer: The car experiences a centripetal force directed towards the center of the circle.)

4. What happens when a force is applied to an object on a rough surface?

(Answer: The object experiences a frictional force opposing the motion.)

5. What happens when a ball is thrown upwards with an initial velocity?

(Answer: The ball experiences a downward acceleration due to gravity.)

6. What happens when a car accelerates from rest to a high speed?

(Answer: The car experiences a forward force due to

the engine's torque.)

7. What happens when a particle moves in a circular path with a constant speed?

(Answer: The particle experiences a centripetal acceleration directed towards the center of the circle.)

8. What happens when a force is applied to an object on an inclined plane?

(Answer: The object experiences a resolved force along the inclined plane.)

9. What happens when a car brakes suddenly to come to rest?

(Answer: The car experiences a deceleration due to the frictional force opposing the motion.)

10. What happens when a projectile is launched at an angle to the horizontal?

(Answer: The projectile experiences a downward acceleration due to gravity and a horizontal velocity component.)

Chapter - 3
Friction: The Hidden Resistance

Friction is an essential force that influences nearly every aspect of motion in our daily lives. It is the resistance that one surface or object encounters when moving over another. Though often overlooked, friction plays a dual role: it can both facilitate motion and create challenges.

How Friction Helps Motion?

1. **Walking and Running**: Without friction between your shoes and the ground, walking or running would be impossible. Static friction prevents your feet from slipping and allows you to push off the ground effectively.
Example: Hiking on a rocky trail relies on strong friction to maintain balance and prevent falls.
2. **Driving**: The grip between tires and the road, known as traction, is crucial for accelerating, turning, and stopping vehicles.

Example: Anti-lock braking systems (ABS) optimize friction to prevent skidding during sudden stops.

3. **Holding Objects**: Friction allows us to grip and hold items securely without them slipping from our hands.

4. **Climbing**: Rock climbers depend on friction between their shoes, hands, and the rock surface to ascend safely.

How Friction Hinders Motion?

1. **Energy Loss**: Friction generates heat, which leads to energy loss. This is evident in machines where moving parts heat up due to friction, reducing efficiency.

Example: Engines require lubrication to minimize friction and prevent overheating.

2. **Wear and Tear**: Continuous friction causes surfaces to degrade over time.

Example: Tires on vehicles wear out due to friction with the road.

3. **Obstacle in Sports**: In games like skiing or ice skating, excessive friction can slow down motion, making it difficult to glide smoothly.

Types of Friction

1. **Static Friction**: Prevents objects from starting to move. It is usually stronger than other types.

Example: A book resting on a table does not move due to static friction.

What is Static Friction?

Static friction is a type of friction that opposes the motion of an object when it is stationary. It is the force that prevents an object from moving when a force is applied to it.

Causes of Static Friction
Static friction is caused by the interaction between two surfaces that are in contact with each other. The surfaces have tiny bumps and irregularities that interlock, causing the surfaces to resist motion.

Mathematical Operation
The force of static friction (Fs) can be calculated using the following equation:
$Fs = \mu_s \times N$
where μ_s is the coefficient of static friction, and N is the normal force (the force perpendicular to the surface).
Examples
1. A book is placed on a table. The force of static friction prevents the book from sliding off the table.
2. A car is parked on a hill. The force of static friction prevents the car from rolling down the hill.

Applications
1. Static friction is used in brakes to stop vehicles.
2. Static friction is used in clutches to transmit power from one shaft to another.

Uses
1. Static friction is used to prevent objects from moving or sliding.
2. Static friction is used to transmit power and motion.
Derivation

The force of static friction can be derived from the concept of surface roughness and the interaction between two surfaces.

Diagram
A diagram showing the force of static friction:

Object---------------Surface

Fs

Conceptual Problems
1. Why does a car need brakes to stop?
2. How does the force of static friction change if the surface roughness is increased?

Coefficient of Static Friction
The coefficient of static friction (μ_s) is a measure of the force of static friction between two surfaces. It is a dimensionless quantity that depends on the surface roughness and the materials in contact.

Types of Static Friction
There are two types of static friction:

a. **Dry static friction**: This type of static friction occurs between two dry surfaces.
b. **Wet static friction**: This type of static friction occurs between two wet surfaces.

Static Friction vs. Kinetic Friction
Static friction is the force that opposes the motion of an object when it is stationary. Kinetic friction, on the

other hand, is the force that opposes the motion of an object when it is moving.

Static Friction Problems
Here are some problems to help you understand static friction:
1. A block of mass 10 kg is placed on a horizontal surface. The coefficient of static friction between the block and the surface is 0.2. What is the maximum force that can be applied to the block without causing it to move?
2. A car is parked on a hill. The coefficient of static friction between the tires and the road is 0.5. What is the minimum force required to get the car moving?

2. **Kinetic (Sliding) Friction**: Acts when objects slide against each other, opposing motion.
Example: Pushing a heavy box across a rough surface.

What is Kinetic or Sliding Friction?
Kinetic or sliding friction is the force that opposes the motion of an object when it is sliding or moving over a surface. It is also known as dynamic friction.

Causes of Kinetic Friction
Kinetic friction is caused by the interaction between two surfaces that are in contact with each other. The surfaces have tiny bumps and irregularities that interlock, causing the surfaces to resist motion.

Mathematical Operation
The force of kinetic friction (F_k) can be calculated using the following equation:

$F_k = \mu_k \times N$

where μ_k is the coefficient of kinetic friction, and N is the normal force (the force perpendicular to the surface).

Applications
1. Kinetic friction is used in brakes to slow down or stop vehicles.
2. Kinetic friction is used in gears to transmit power and motion.

Uses
1. Kinetic friction is used to slow down or stop objects in motion.
2. Kinetic friction is used to transmit power and motion.

Derivation
The force of kinetic friction can be derived from the concept of surface roughness and the interaction between two surfaces.

Examples
1. A block of wood is sliding over a rough surface. The force of kinetic friction opposes the motion of the block.
2. A car is moving over a wet road. The force of kinetic friction opposes the motion of the car.

Diagram
A diagram showing the force of kinetic friction:

Object---------------Surface
F_k

Conceptual Problems

1. Why does a car need brakes to slow down or stop?
2. How does the force of kinetic friction change if the surface roughness is increased?

Coefficient of Kinetic Friction

The coefficient of kinetic friction (μ_k) is a measure of the force of kinetic friction between two surfaces. It is a dimensionless quantity that depends on the surface roughness and the materials in contact.

Types of Kinetic Friction

There are two types of kinetic friction:

1. **Dry kinetic friction**: This type of kinetic friction occurs between two dry surfaces.
2. **Wet kinetic friction**: This type of kinetic friction occurs between two wet surfaces.

Kinetic Friction vs. Static Friction

Kinetic friction is the force that opposes the motion of an object when it is sliding or moving over a surface. Static friction, on the other hand, is the force that opposes the motion of an object when it is stationary.

Kinetic Friction Problems

Here are some problems to help you understand kinetic friction:

1. A block of mass 10 kg is sliding over a rough surface. The coefficient of kinetic friction between the block and the surface is 0.2. What is the force of kinetic friction opposing the motion of the block?
2. A car is moving over a wet road. The coefficient of kinetic friction between the tires and the road is 0.5. What is the force of kinetic friction opposing the

motion of the car?

3. **Rolling Friction**: Occurs when objects roll over a surface, like wheels on a road. It is weaker than sliding friction, which is why vehicles use wheels.
Example: Bicycles move efficiently because rolling friction is minimized.

What is Rolling Friction?
Rolling friction is the force that opposes the motion of an object that is rolling over a surface. It is also known as rolling resistance.

Causes of Rolling Friction
Rolling friction is caused by the interaction between the rolling object and the surface it is rolling on. The surface has tiny bumps and irregularities that cause the rolling object to experience a force opposing its motion.

Mathematical Operation
The force of rolling friction (F_r) can be calculated using the following equation:
$F_r = \mu_r \times N$
where μ_r is the coefficient of rolling friction, and N is the normal force (the force perpendicular to the surface).

Examples

1. A wheelbarrow is rolling over a rough surface. The force of rolling friction opposes the motion of the wheelbarrow.
2. A car is moving over a smooth road. The force of rolling friction opposes the motion of the car.

Applications

1. Rolling friction is used in the design of wheels and bearings to reduce the force opposing motion.
2. Rolling friction is used in the design of gears and transmissions to transmit power and motion.

Uses
1. Rolling friction is used to slow down or stop objects in motion.
2. Rolling friction is used to transmit power and motion.

Derivation
The force of rolling friction can be derived from the concept of surface roughness and the interaction between the rolling object and the surface.

Diagram
A diagram showing the force of rolling friction:

Object---------------Surface
F_r

Conceptual Problems
1. Why does a wheelbarrow experience a force opposing its motion when it is rolling over a rough surface?
2. How does the force of rolling friction change if the surface roughness is increased?

Coefficient of Rolling Friction
The coefficient of rolling friction (μ_r) is a measure of the force of rolling friction between two surfaces. It is a dimensionless quantity that depends on the surface

roughness and the materials in contact.

Types of Rolling Friction
There are two types of rolling friction:
1. **Dry rolling friction**: This type of rolling friction occurs between two dry surfaces.
2. **Wet rolling friction**: This type of rolling friction occurs between two wet surfaces.

Rolling Friction vs. Sliding Friction
Rolling friction is the force that opposes the motion of an object that is rolling over a surface. Sliding friction, on the other hand, is the force that opposes the motion of an object that is sliding over a surface.

Rolling Friction Problems
Here are some problems to help you understand rolling friction:
1. A wheelbarrow is rolling over a rough surface. The coefficient of rolling friction between the wheelbarrow and the surface is 0.1. What is the force of rolling friction opposing the motion of the wheelbarrow?
2. A car is moving over a smooth road. The coefficient of rolling friction between the tires and the road is 0.05. What is the force of rolling friction opposing the motion of the car?

What are Pulleys?
A pulley is a wheel with a grooved rim and a rope, cable, or chain wrapped around it. Pulleys are used to change the direction of force or motion, or to gain a mechanical advantage in lifting or moving heavy loads.

Types of Pulleys
There are several types of pulleys, including:
1. **Fixed Pulley**: A fixed pulley is attached to a fixed point, such as a beam or a hook. It changes the direction of the force or motion, but does not provide a mechanical advantage.
2. **Movable Pulley**: A movable pulley is attached to the load being lifted or moved. It provides a mechanical advantage, as the force applied to the rope or cable is multiplied by the number of ropes or cables supporting the load.
3. **Compound Pulley**: A compound pulley is a combination of fixed and movable pulleys. It provides a greater mechanical advantage than a single pulley, as the force applied to the rope or cable is multiplied by the number of ropes or cables supporting the load.
4. **Block and Tackle Pulley**: A block and tackle pulley is a type of compound pulley that uses a system of ropes and pulleys to provide a mechanical advantage.

How Pulleys Decrease Friction?
Pulleys decrease friction in several ways:
1. **Reducing the force required to lift or move a load**: By providing a mechanical advantage, pulleys reduce the force required to lift or move a load, which reduces the frictional force opposing the motion.
2. **Changing the direction of the force or motion**: Pulleys change the direction of the force or motion, which reduces the frictional force opposing the motion.
3. **Using a smooth surface**: Pulleys use a smooth surface, such as a wheel or a groove, to reduce the frictional force opposing the motion.

Mathematical Operations

Here are some mathematical operations involved with pulleys:

1. **Mechanical Advantage**: The mechanical advantage of a pulley system is the ratio of the output force to the input force. It can be calculated using the following equation:

$MA = F_{out} / F_{in}$

where MA is the mechanical advantage, F_{out} is the output force, and F_{in} is the input force.

2. **Efficiency**: The efficiency of a pulley system is the ratio of the output work to the input work. It can be calculated using the following equation:

$\eta = W_{out} / W_{in}$

where η is the efficiency, W_{out} is the output work, and W_{in} is the input work.

3. **Force and Motion**: The force and motion of a pulley system can be calculated using the following equations:

$F_{out} = F_{in} \times MA$

$v_{out} = v_{in} / MA$

where F_{out} is the output force, F_{in} is the input force, v_{out} is the output velocity, and v_{in} is the input velocity.

Solutions and Concepts Involved with Pulleys

Here are some solutions and concepts involved with pulleys:

1. **Lifting a Heavy Load**: A pulley system can be used to lift a heavy load by providing a mechanical advantage. For example, a block and tackle pulley system can be used to lift a heavy load by providing a mechanical advantage of 3:1 or 4:1.

2. **Changing the Direction of a Force**: A pulley system can be used to change the direction of a force. For example, a fixed pulley can be used to change the direction of a force from horizontal to vertical.

3. **Reducing Friction**: A pulley system can be used to reduce friction. For example, a pulley system with a smooth surface can be used to reduce the frictional force opposing the motion.

Here are the answers to the conceptual problems:

1. What is the mechanical advantage of a pulley system?

The mechanical advantage of a pulley system is the ratio of the output force to the input force. It is a measure of the amount of force amplification provided by the pulley system.

2. How does a pulley system change the direction of a force?

A pulley system changes the direction of a force by using a wheel or a grooved pulley to redirect the force. For example, a fixed pulley can change the direction of a force from horizontal to vertical.

3. What is the efficiency of a pulley system?

The efficiency of a pulley system is the ratio of the output work to the input work. It is a measure of the amount of energy lost as heat or friction in the pulley system.

4. How does a pulley system reduce friction?

A pulley system reduces friction by using a smooth surface, such as a wheel or a grooved pulley, to reduce the frictional force opposing the motion.

5. What are the different types of pulleys, and how are they used?

There are several types of pulleys, including fixed pulleys, movable pulleys, compound pulleys, and block and tackle pulleys. Each type of pulley is used for a specific application, such as lifting heavy loads, changing the direction of a force, or providing a mechanical advantage.

Reducing friction and protecting moving parts from wear and tear are crucial in maintaining the efficiency and longevity of mechanical systems. Here are some ways to achieve this:

Methods to Reduce Friction

1. **Lubrication**: Apply lubricants such as oils, greases, or silicone sprays to reduce friction between moving parts.
2. **Bearings**: Use ball bearings, roller bearings, or needle bearings to reduce friction and support loads.
3. **Bushings**: Install bushings or sleeves to reduce friction and wear on moving parts.
4. **Surface Finishing**: Apply surface finishing techniques such as polishing, grinding, or honing to reduce friction and wear.
5. **Materials Selection**: Choose materials with low friction coefficients, such as Teflon, nylon, or polyethylene.

Methods to Protect Moving Parts

1. **Seals and Gaskets**: Use seals and gaskets to prevent lubricants from escaping and contaminants from entering.
2. **Protective Coatings**: Apply protective coatings such as paint, varnish, or powder coating to prevent corrosion and wear.

3. **Corrosion-Resistant Materials**: Use corrosion-resistant materials such as stainless steel, aluminum, or titanium.
4. **Regular Maintenance**: Perform regular maintenance tasks such as cleaning, lubricating, and inspecting moving parts.
5. **Design for Wear**: Design moving parts with wear-resistant features such as rounded edges, smooth surfaces, and generous clearances.

Additional Tips
1. **Monitor Temperature**: Monitor temperature to prevent overheating, which can increase friction and wear.
2. **Use the Right Lubricant**: Use the right lubricant for the application, taking into account factors such as temperature, load, and speed.
3. **Avoid Over-Tightening**: Avoid over-tightening, which can increase friction and wear.
4. **Use Wear-Resistant Materials**: Use wear-resistant materials such as hardened steel, ceramic, or tungsten carbide.
5. **Design for Easy Maintenance**: Design moving parts with easy maintenance in mind, such as easy access to lubrication points.

Force of Friction in a Lift Going Upward

When a lift is going upward, the force of friction acts in the opposite direction to the motion of the lift. This is because the force of friction is opposing the motion of the lift.

Mathematically, the force of friction (F_f) can be

calculated using the following equation:
$F_f = \mu \times N$
where μ is the coefficient of friction, and N is the normal force (the force perpendicular to the surface).
In the case of a lift going upward, the normal force (N) is equal to the weight of the lift (W) plus the force applied to lift the load ($F_{applied}$):
$N = W + F_{applied}$
The force of friction (F_f) can then be calculated as:
$F_f = \mu \times (W + F_{applied})$

Force of Friction in a Lift Going Downward

When a lift is going downward, the force of friction acts in the same direction as the motion of the lift. This is because the force of friction is opposing the downward motion of the lift.
Mathematically, the force of friction (F_f) can be calculated using the following equation:
$F_f = \mu \times N$
where μ is the coefficient of friction, and N is the normal force (the force perpendicular to the surface).
In the case of a lift going downward, the normal force (N) is equal to the weight of the lift (W) minus the force applied to control the descent of the load ($F_{applied}$):
$N = W - F_{applied}$
The force of friction (F_f) can then be calculated as:
$F_f = \mu \times (W - F_{applied})$

Example Problem
A lift is moving upward with a load of 1000 kg. The coefficient of friction between the lift and the guide rails is 0.1. If the force applied to lift the load is 5000

N, calculate the force of friction opposing the motion of the lift.

Solution:
Weight of the load (W) = 1000 kg x 9.8 m/s² = 9800 N
Normal force (N) = W + F$_{applied}$ = 9800 N + 5000 N = 14800 N
Force of friction (F$_f$) = μ x N = 0.1 x 14800 N = 1480 N

Therefore, the force of friction opposing the motion of the lift is 1480 N.

Balancing Friction in Practical Applications

1. Engineering: Machines are designed to minimize friction in moving parts with lubricants or ball bearings to enhance efficiency.
2. Sports Equipment: Designers optimize friction for specific needs, like high-friction basketball shoes for better grip or low-friction ice skates for speed.
3. Transportation: Roads are engineered with textures that provide adequate friction for safe driving.
4. The role of friction in motion and how it can both help and hinder.

Conceptual Insights

Friction is a force that demands respect and understanding. Without it, motion would be chaotic and uncontrollable. Conversely, excessive friction can slow progress or cause damage. Engineers, scientists, and everyday individuals continually seek to manage friction to harness its benefits while mitigating its drawbacks.

By studying friction, students can appreciate the intricate balance required in designing systems and solving real-world problems, from creating energy-efficient machines to ensuring road safety.

Here are the solutions to the NCERT questions from the exercises of Class XI Physics concerning friction:

1. Define friction.

Answer: Friction is the force that opposes the motion of an object when it is in contact with another surface.

2. List the factors on which the friction depends.

Answer: The friction depends on the following factors:

- Nature of the surfaces in contact

- Area of contact between the surfaces

- Normal force acting between the surfaces

3. Discuss the laws of friction.

Answer: The laws of friction are:

- The force of friction is directly proportional to the normal force acting between the surfaces.

- The force of friction is independent of the area of contact between the surfaces.

- The force of friction depends on the nature of the surfaces in contact.

4. The angle of contact between two surfaces is 30°. If the coefficient of friction is 0.5, calculate the angle

of friction.

Answer: Angle of friction = $\tan^{-1}(\mu) = \tan^{-1}(0.5) = 26.57°$

5 A block of mass 2 kg is placed on a horizontal surface. The coefficient of kinetic friction between the block and the surface is 0.1. If a horizontal force of 10 N is applied to the block, calculate the acceleration of the block.

Answer: Force of friction = $\mu \times N = 0.1 \times 2 \times 9.8 = 1.96$ N

Net force acting on the block = 10 - 1.96 = 8.04 N

Acceleration of the block = Net force / Mass = 8.04 / 2 = 4.02 m/s^2

6. A car is moving on a horizontal road with a uniform speed of 30 km/h. If the coefficient of kinetic friction between the car and the road is 0.03, calculate the power of the engine.

Answer: Force of friction = $\mu \times N = 0.03 \times 1500 \times 9.8 = 441.5$ N

Power of the engine = Force x Velocity = 441.5 x (30 x 1000 / 3600) = 11040 W

7. A block of mass 5 kg is placed on a horizontal surface. The coefficient of static friction between the block and the surface is 0.2. If a horizontal force of 10 N is applied to the block, calculate the force of friction acting on the block.

Answer: Force of friction = $\mu \times N = 0.2 \times 5 \times 9.8 = 9.8$ N

Since the applied force (10 N) is greater than the force of friction (9.8 N), the block will move.

8. A car is moving on a horizontal road with a uniform speed of 40 km/h. If the coefficient of kinetic friction between the car and the road is 0.05, calculate the force of friction acting on the car.

Answer: Force of friction = $\mu \times N$ = 0.05 x 2000 x 9.8 = 980 N

9. A block of mass 10 kg is placed on an inclined plane with an angle of 30°. The coefficient of kinetic friction between the block and the plane is 0.1. If the block is released from rest, calculate the acceleration of the block.

Answer: Force of friction = $\mu \times N$ = 0.1 x 10 x 9.8 x cos (30°) = 8.49 N

Net force acting on the block = mg x sin (30°) - 8.49 = 49 - 8.49 = 40.51 N

Acceleration of the block = Net force / Mass = 40.51 / 10 = 4.051 m/s^2

Here are twenty "what happens when" questions with answers related to friction for Class XI Physics:

1. What happens when you try to slide a heavy box across a rough floor?

Answer: The box will experience a large force of friction, making it difficult to slide.

2. What happens when you apply brakes to a moving car?

Answer: The force of friction between the brake pads and the wheels slows down the car.

3. What happens when you try to walk on a slippery

floor?

Answer: You will experience a low force of friction, making it difficult to walk without slipping.

4. What happens when you use a lubricant on a squeaky door hinge?

Answer: The lubricant reduces the force of friction between the hinge and the door, making it easier to open and close.

5. What happens when you try to push a heavy object up a rough inclined plane?

Answer: The force of friction will oppose the motion of the object, making it more difficult to push up the plane.

6. What happens when you use a rope with a rough surface to pull a load?

Answer: The force of friction between the rope and the load will be higher, making it more difficult to pull the load.

7 What happens when you try to stop a moving bicycle by applying the brakes?

Answer: The force of friction between the brake pads and the wheels slows down the bicycle.

8. What happens when you use a bearing to support a rotating shaft?

Answer: The bearing reduces the force of friction between the shaft and the support, making it easier to rotate the shaft.

9. What happens when you try to pull a heavy load across a smooth floor?

Answer: The force of friction will be lower, making it easier to pull the load.

10. What happens when you use a lubricant on a machine?

Answer: The lubricant reduces the force of friction between the moving parts, making it easier to operate the machine.

11. What happens when you try to walk on a rough surface with smooth-soled shoes?

Answer: You will experience a lower force of friction, making it more difficult to walk without slipping.

12. What happens when you use a rope with a smooth surface to pull a load?

Answer: The force of friction between the rope and the load will be lower, making it easier to pull the load.

13. What happens when you try to push a heavy object down a rough inclined plane?

Answer: The force of friction will oppose the motion of the object, making it more difficult to push down the plane.

14. What happens when you use a bearing to support a rotating shaft in a machine?

Answer: The bearing reduces the force of friction between the shaft and the support, making it easier to rotate the shaft and increasing the efficiency of the machine.

15. What happens when you try to stop a moving car

by applying the brakes on a slippery road?

Answer: The force of friction between the brake pads and the wheels will be lower, making it more difficult to stop the car.

16. What happens when you use a lubricant on a machine that is subject to high temperatures?

Answer: The lubricant will reduce the force of friction between the moving parts, but it may also break down due to the high temperatures, reducing its effectiveness.

17. What happens when you try to pull a heavy load across a rough surface with a rope that has a high coefficient of friction?

Answer: The force of friction between the rope and the load will be higher, making it more difficult to pull the load.

18. What happens when you use a bearing to support a rotating shaft in a machine that is subject to high loads?

Answer: The bearing will reduce the force of friction between the shaft and the support, making it easier to rotate the shaft and increasing the efficiency of the machine.

19. What happens when you try to stop a moving bicycle by applying the brakes on a steep downhill slope?

Answer: The force of friction between the brake pads and the wheels will be lower, making it more difficult to stop the bicycle.

20. What happens when you use a lubricant on a

machine that is subject to high vibrations?

Answer: The lubricant will reduce the force of friction between the moving parts, but it may also be displaced due to the high vibrations, reducing its effectiveness.

Here's an exercise with hints for Class XI Physics on the topics of friction, pulleys, and inclined planes:

Very Short Answer Type Questions (1 mark each)

1. What is the force that opposes the motion of an object?

(Hint: Friction)

2. What is the coefficient of friction?

(Hint: Ratio of frictional force to normal force)

3. What is the purpose of a pulley?

(Hint: Change direction of force or motion)

4. What is the difference between static and kinetic friction?

(Hint: Static friction prevents motion, kinetic friction opposes motion)

5. What is the angle of friction?

(Hint: Angle between normal force and resultant force)

6. What is the force that opposes the motion of an object on an inclined plane?

(Hint: Frictional force)

7. What is the purpose of a lubricant?

(Hint: Reduce friction)

8. What is the difference between a fixed pulley and a movable pulley?

(Hint: Fixed pulley changes direction of force, movable pulley changes magnitude of force)

9. What is the coefficient of kinetic friction?

(Hint: Ratio of frictional force to normal force during motion)

10. What is the purpose of a bearing?

(Hint: Reduce friction and wear)

Short Answer Type Questions (3 marks each)

1. Describe the different types of friction. (Hint: Static, kinetic, and rolling friction)

2. Explain the concept of coefficient of friction. (Hint: Ratio of frictional force to normal force)

3. Describe the working of a pulley system. (Hint: Change direction of force or motion)

4. Explain the concept of angle of friction. (Hint: Angle between normal force and resultant force)

5. Describe the factors that affect the frictional force. (Hint: Nature of surfaces, area of contact, normal force)

6. Explain the concept of lubrication. (Hint: Reduce friction using lubricants)

7. Describe the different types of bearings. (Hint: Ball bearings, roller bearings, and needle bearings)

8. Explain the concept of efficiency of a pulley system. (Hint: Ratio of output work to input work)

9. Describe the factors that affect the efficiency of a pulley system. (Hint: Friction, weight of the load, and mechanical advantage)

10. Explain the concept of mechanical advantage of a pulley system. (Hint: Ratio of output force to input force)

Long Questions (5 marks each)

1. Describe the different types of friction and explain the factors that affect the frictional force. (Hint: Static, kinetic, and rolling friction, and factors such as nature of surfaces, area of contact, and normal force)

2. Explain the concept of coefficient of friction and describe its significance in engineering applications. (Hint: Ratio of frictional force to normal force, and significance in designing machines and mechanisms)

3. Describe the working of a pulley system and explain the concept of mechanical advantage. (Hint: Change direction of force or motion, and ratio of output force to input force)

4. Explain the concept of lubrication and describe its significance in reducing friction and wear. (Hint: Reduce friction using lubricants, and significance in increasing efficiency and reducing wear)

5. Describe the different types of bearings and explain their significance in reducing friction and wear. (Hint: Ball bearings, roller bearings, and needle bearings, and significance in increasing efficiency and reducing wear)

Numerical Questions (5 marks each)

1. A block of mass 10 kg is placed on a horizontal surface. The coefficient of kinetic friction between the block and the surface is 0.2. If a horizontal force of 50 N is applied to the block, calculate the acceleration of the block.

2. A pulley system has a mechanical advantage of 3. If a force of 100 N is applied to the input rope, calculate the force exerted on the load.

3. A block of mass 20 kg is placed on an inclined plane with an angle of 30°. The coefficient of kinetic friction between the block and the plane is 0.1. If the block is released from rest, calculate the acceleration of the block.

4. A wheel of radius 0.5 m is rotating with an angular velocity of 10 rad/s. If the coefficient of kinetic friction between the wheel and the ground is 0.2, calculate the torque required to maintain the rotation.

5. A rope is wrapped around a cylindrical drum of radius 0.2 m. If a force of 50 N is applied to the rope, calculate the torque exerted on the drum.

Chapter - 4
Circular Motion and Centripetal Force

Physics is replete with fascinating phenomena that govern the natural world, and one of the most intriguing is circular motion. Circular motion refers to the movement of an object along the circumference of a circle. Whether it's the planets orbiting the Sun, a car navigating a curve, or electrons whizzing around the nucleus of an atom, circular motion is fundamental to our understanding of rotational dynamics.

Circular Motion
Circular motion occurs when an object moves along a circular path. This motion can be classified into two types:

1. **Uniform Circular Motion (UCM):**
The object moves with constant speed along a circular path. Even though the speed is constant, the direction of velocity changes continuously, leading to acceleration.
Example: The hands of a clock.

Uniform circular motion is a type of circular motion where the speed of the object remains constant, but the direction of the velocity changes continuously. This type of motion is common in everyday life, from the rotation of the Earth to the spinning of a top.

Characteristics of Uniform Circular Motion
The following are the key characteristics of uniform circular motion:

a. **Constant Speed**: The speed of the object remains constant throughout the motion.
b. **Changing Direction**: The direction of the velocity changes continuously as the object moves in a circular path.
c. **Constant Acceleration**: The acceleration of the object is constant in magnitude, but changes in direction continuously.
d. **Centripetal Force**: A centripetal force is required to keep the object moving in a circular path.

Mathematical Operations
The following are the key mathematical operations used to describe uniform circular motion:
1. **Centripetal Force**: The centripetal force (F_c) can be calculated using the following equation:
$F_c = (m \times v^2) / r$
where:
m = mass of the object
v = velocity of the object
r = radius of the circle
2. **Centripetal Acceleration**: The centripetal acceleration (a_c) can be calculated using the following

equation:
$A_c = v^2 / r$
where:
v = velocity of the object
r = radius of the circle

3 **Angular Velocity**: The angular velocity (ω) can be calculated using the following equation:
$\omega = v / r$
where:
v = velocity of the object
r = radius of the circle

Applications and Uses

Uniform circular motion has many applications and uses in everyday life, including:

1. **Transportation**: Cars, buses, and trains all use uniform circular motion to move.
2. **Astronomy**: The Earth and other planets orbit the Sun in uniform circular motion.
3. **Engineering**: Uniform circular motion is used in the design of gears, pulleys, and other mechanical systems.
4. **Sports**: Many sports, such as cycling, skating, and gymnastics, involve uniform circular motion.

Conceptual Problems

Here are some conceptual problems to help you understand uniform circular motion:

1. What is the direction of the centripetal force?
2. What is the relationship between the velocity and the radius of a circular path?
3. What is the difference between uniform and non-uniform circular motion?
4. How does the centripetal force change as the velocity of an object increases?

5. What is the role of friction in uniform circular motion?

Numerical Problems

Here are some numerical problems to help you practice uniform circular motion:

1. A car is moving in a circular path with a radius of 50 m. If the velocity of the car is 20 m/s, calculate the centripetal force acting on the car.

2. A bicycle is moving in a circular path with a radius of 10 m. If the velocity of the bicycle is 15 m/s, calculate the centripetal acceleration acting on the bicycle.

3. A planet is orbiting the Sun in a circular path with a radius of 1.5×10^{11} m. If the velocity of the planet is 30 km/s, calculate the centripetal force acting on the planet.

4. A car is moving in a circular path with a radius of 20 m. If the velocity of the car is 25 m/s, calculate the angular velocity of the car.

5. A bicycle is moving in a circular path with a radius of 5 m. If the velocity of the bicycle is 10 m/s,

2. **Non-Uniform Circular Motion:**

The speed of the object changes as it moves along the circular path.

Example: A car accelerating while taking a turn.

Non-uniform circular motion is a type of circular motion where the speed of the object changes as it moves in a circular path. This type of motion is common in everyday life, from the rotation of a merry-go-round to the orbit of a planet around the Sun.

Characteristics of Non-Uniform Circular Motion
The following are the key characteristics of non-uniform circular motion:
1. **Changing Speed**: The speed of the object changes as it moves in a circular path.
2. **Changing Direction**: The direction of the velocity changes continuously as the object moves in a circular path.
3. **Tangential Acceleration**: The object experiences a tangential acceleration, which is the rate of change of the speed.
4. **Centripetal Acceleration**: The object also experiences a centripetal acceleration, which is the rate of change of the direction.

Mathematical Operations
The following are the key mathematical operations used to describe non-uniform circular motion:
1. **Tangential Acceleration**: The tangential acceleration (a_t) can be calculated using the following equation:
$a_t = \Delta v / \Delta t$
where:
Δv = change in velocity
Δt = time interval
2. **Centripetal Acceleration**: The centripetal acceleration (a_c) can be calculated using the following equation:
$a_c = v^2 / r$
where:
v = velocity of the object
r = radius of the circle
3. **Total Acceleration**: The total acceleration (a) can be

calculated using the following equation:
$a = \sqrt{(a_t^2 + a_c^2)}$
where:
a_t = tangential acceleration
a_c = centripetal acceleration

Here are the expressions for tangential acceleration and centripetal acceleration using calculus:

Tangential Acceleration

The tangential acceleration (a_t) is the rate of change of the tangential velocity (v_t). Using calculus, we can express the tangential acceleration as:

$a_t = dv_t / dt$

where:
dv_t = change in tangential velocity
dt = time interval

Centripetal Acceleration

The centripetal acceleration (a_c) is the rate of change of the centripetal velocity (v_c). Using calculus, we can express the centripetal acceleration as:

$a_c = dv_c / dt$

where:
dv_c = change in centripetal velocity
dt = time interval

However, since the centripetal velocity is perpendicular to the radius vector, we can use the following expression:
$a_c = -v^2 / r$
where:
v = velocity of the object
r = radius of the circle

The negative sign indicates that the centripetal acceleration is directed towards the center of the circle.

Total Acceleration

The total acceleration (a) of an object moving in a circular path is the vector sum of the tangential acceleration and the centripetal acceleration:

$$a = \sqrt{(a_t^2 + a_c^2)}$$

Using calculus, we can express the total acceleration as:

$$a = \sqrt{((dv_t / dt)^2 + (v^2 / r^2))}$$

where:
dv_t = change in tangential velocity
dt = time interval
v = velocity of the object
r = radius of the circle

4. Angular Displacement

Angular displacement is the change in the angular position of an object. It is a measure of the amount of rotation of an object.

Mathematical Operations:

The angular displacement (θ) can be calculated using the following equation:

$$\theta = \theta_f - \theta_i$$

where:
θ_f = final angular position
θ_i = initial angular position

The angular displacement can also be calculated using the following equation:

$$\theta = s / r$$

where:
s = arc length
r = radius of the circle

5. Angular Velocity

Angular velocity is the rate of change of the angular

displacement. It is a measure of the speed of rotation of an object.

Mathematical Operations:

The angular velocity (ω) can be calculated using the following equation:

$\omega = \Delta\theta / \Delta t$

where:

$\Delta\theta$ = change in angular displacement

Δt = time interval

The angular velocity can also be calculated using the following equation:

$\omega = v / r$

where:

v = linear velocity

r = radius of the circle

Relationship between Angular Displacement and Angular Velocity

The angular displacement and angular velocity are related by the following equation:

$\omega = d\theta / dt$

where:

ω = angular velocity

θ = angular displacement

t = time

This equation shows that the angular velocity is the rate of change of the angular displacement.

Units

The units of angular displacement and angular velocity are:

Angular Displacement: radians (rad)

Angular Velocity: radians per second (rad/s)

Applications and Uses
Non-uniform circular motion has many applications and uses in everyday life, including:
1. **Transportation**: Cars, buses, and trains all use non-uniform circular motion to move.
2. **Astronomy**: The orbit of a planet around the Sun is an example of non-uniform circular motion.
3. **Engineering**: Non-uniform circular motion is used in the design of gears, pulleys, and other mechanical systems.
4. **Sports**: Many sports, such as cycling, skating, and gymnastics, involve non-uniform circular motion.

Centripetal Force

Centripetal force is the net force required to make an object follow a circular path. Unlike forces like tension, gravity, or friction, centripetal force is not a separate force but rather the resultant force providing the centripetal acceleration.

Centripetal force is a force that acts on an object moving in a circular path, directed towards the center of the circle. This force is necessary to keep the object moving in a circular path, and it is responsible for the change in direction of the object's velocity.

Characteristics of Centripetal Force
The following are the key characteristics of centripetal force:
1. **Direction**: Centripetal force acts towards the center of the circle.
2. **Magnitude**: The magnitude of the centripetal force

depends on the mass of the object, the velocity of the object, and the radius of the circle.

3. **Type of force**: Centripetal force is a contact force, meaning it acts through direct contact between objects.

Mathematical Operations

The centripetal force (F_c) can be calculated using the following equation:

$$F_c = (m \times v^2) / r$$

where:

m = mass of the object
v = velocity of the object
r = radius of the circle

Examples of Centripetal Force

Applications and Uses

Centripetal force has many applications and uses in everyday life, including:

1. **Transportation**: Cars, buses, and trains all rely on centripetal force to move around curves.

2. **Astronomy**: The orbits of planets around the Sun are due to centripetal force.

3. **Engineering**: Centripetal force is used in the design of circular structures, such as bridges and tunnels.

4. **Sports**: Many sports, such as cycling and skating, involve centripetal force.

Conceptual Problems

Here are some conceptual problems to help you understand centripetal force:

1. What is the direction of the centripetal force?

2. What is the relationship between the velocity and the radius of a circular path?

3. What is the difference between centripetal force and

centrifugal force?
4. How does the centripetal force change as the velocity of an object increases?
5. What is the role of friction in centripetal force?

Examples
1. **Satellites in Orbit:**

The gravitational force acts as the centripetal force, keeping the satellite in a stable orbit.

Concept
When a satellite orbits a planet, it is constantly falling towards the planet due to the gravitational force between them. However, because the satellite is also moving tangentially to the planet, it never actually gets closer to the planet. Instead, it follows a curved path around the planet, which is its orbit.

The gravitational force between the satellite and the planet acts as the centripetal force, keeping the satellite in its orbit. The centripetal force is directed towards the center of the planet, and it is responsible for keeping the satellite moving in a curved path.

Mathematical Derivation
Let's consider a satellite of mass m orbiting a planet of mass M. The gravitational force between the satellite and the planet is given by:
$F_g = G \times (m \times M) / r^2$
where:
F_g = gravitational force
G = gravitational constant
m = mass of the satellite
M = mass of the planet

r = distance between the satellite and the planet
Since the gravitational force acts as the centripetal force, we can set it equal to the centripetal force:
$F_g = F_c$
$F_c = (m \times v^2) / r$
where:

F_c = centripetal force
m = mass of the satellite
v = velocity of the satellite
r = radius of the orbit
Equating the two expressions for the gravitational force and the centripetal force, we get:
$G \times (m \times M) / r^2 = (m \times v^2) / r$
Simplifying and rearranging, we get:
$V^2 = G \times M / r$
This is the equation for the velocity of a satellite in a stable orbit around a planet. It shows that the velocity of the satellite depends on the mass of the planet, the radius of the orbit, and the gravitational constant.

Orbital Period
The orbital period (T) of a satellite is the time it takes to complete one orbit around the planet. We can derive an expression for the orbital period using the equation for the velocity of the satellite:
$T = 2 \times \pi \times r / v$
Substituting the expression for the velocity, we get:
$T = 2 \times \pi \times r / \sqrt{(G \times M / r)}$
Simplifying and rearranging, we get:
$T = 2 \times \pi \times \sqrt{(r^3 / (G \times M))}$
This is the equation for the orbital period of a satellite in a stable orbit around a planet. It shows that the

orbital period depends on the radius of the orbit, the mass of the planet, and the gravitational constant.

In conclusion, the gravitational force acts as the centripetal force keeping a satellite in a stable orbit around a planet. The velocity of the satellite depends on the mass of the planet, the radius of the orbit, and the gravitational constant. The orbital period of the satellite depends on the radius of the orbit, the mass of the planet, and the gravitational constant.

2. Car Turning on a Curve:

Friction between the tires and the road provides the centripetal force required for the turn.

Concept

When a vehicle turns, it follows a curved path. To maintain this curved path, a force is required to keep the vehicle moving in a circle. This force is provided by the friction between the tires and the road.

The frictional force acts as the centripetal force, directing the vehicle towards the center of the turn. The magnitude of the frictional force depends on the coefficient of friction between the tires and the road, as well as the normal force exerted by the road on the vehicle.

Mathematical Derivation

Let's consider a vehicle of mass m moving with a velocity v in a circular path of radius r. The frictional force (F_f) acts as the centripetal force, keeping the vehicle moving in a circle.

The frictional force can be calculated using the following equation:

$F_f = \mu \times N$

where:
F_f = frictional force
μ = coefficient of friction
N = normal force

The normal force (N) is equal to the weight of the vehicle (mg), where g is the acceleration due to gravity.
$N = mg$
Substituting this expression into the equation for the frictional force, we get:
$F_f = \mu \times mg$
Since the frictional force acts as the centripetal force, we can set it equal to the centripetal force:
$F_f = F_c$
$F_c = (m \times v^2) / r$
where:
F_c = centripetal force
m = mass of the vehicle
v = velocity of the vehicle
r = radius of the turn
Equating the two expressions for the frictional force and the centripetal force, we get:
$\mu \times mg = (m \times v^2) / r$
Simplifying and rearranging, we get:
$V^2 = \mu \times g \times r$
This is the equation for the velocity of a vehicle in a turn, in terms of the coefficient of friction, the acceleration due to gravity, and the radius of the turn.

Maximum Velocity

The maximum velocity (v_{max}) that a vehicle can achieve in a turn is limited by the coefficient of friction between the tires and the road. If the vehicle exceeds this velocity, it will skid or lose traction.

On a banked road, the normal reaction and friction together provide this force.

Banking of Road
Banking of a road is a technique used to design roads that curve or bend. The road is tilted or inclined at an angle to ensure that the force of gravity acting on a vehicle is balanced by the force of friction, allowing the vehicle to move smoothly around the curve without skidding or losing traction.

Mathematical Derivation
Let's consider a vehicle of mass m moving with a velocity v around a curved road with a radius of curvature r. The road is banked at an angle θ to the horizontal.

The force of gravity acting on the vehicle is balanced by the normal reaction force (N) exerted by the road on the vehicle. The normal reaction force can be resolved into two components: a horizontal component ($N \sin \theta$) that acts towards the center of the curve, and a vertical component ($N \cos \theta$) that acts upwards.

The frictional force (f) acts in the opposite direction to the motion of the vehicle, and can be calculated using the following equation:

$f = \mu \times N$

where μ is the coefficient of friction between the tires and the road.

The centripetal force (F_c) required to keep the vehicle moving in a circular path can be calculated using the

following equation:
$$F_c = (m \times v^2) / r$$
Since the normal reaction force and the frictional force together provide the centripetal force, we can set up the following equation:
$$N \sin \theta + f = (m \times v^2) / r$$
Substituting the expression for the frictional force, we get:
$$N \sin \theta + \mu \times N = (m \times v^2) / r$$
Simplifying and rearranging, we get:
$$N (\sin \theta + \mu) = (m \times v^2) / r$$
Dividing both sides by $(\sin \theta + \mu)$, we get:
$$\mathbf{N = (m \times v^2) / (r \times (\sin \theta + \mu))}$$
This is the equation for the normal reaction force exerted by the road on the vehicle.

Banked Road

On a banked road, the normal reaction force and the frictional force together provide the centripetal force required to keep the vehicle moving in a circular path. The angle of banking (θ) can be calculated using the following equation:
$$\mathbf{\tan \theta = v^2 / (r \times g)}$$
where g is the acceleration due to gravity.

This equation shows that the angle of banking depends on the velocity of the vehicle, the radius of the curve, and the acceleration due to gravity.

By banking the road at the correct angle, the force of gravity acting on the vehicle is balanced by the force of friction, allowing the vehicle to move smoothly around the curve without skidding or losing traction.

3. Planetary Orbits:

Planets orbit the Sun in elliptical paths, where gravity acts as the centripetal force.

Planetary Orbits

Planetary orbits refer to the paths that planets follow as they revolve around the Sun. The orbits of the planets are elliptical in shape, meaning that they are not perfect circles. The Sun is located at one of the two foci of the elliptical orbit.

Gravity as the Centripetal Force

The force of gravity between the Sun and a planet acts as the centripetal force that keeps the planet in its orbit. The centripetal force is directed towards the center of the Sun, and it is responsible for keeping the planet moving in a curved path.

Mathematical Derivation

Let's consider a planet of mass m orbiting the Sun of mass M. The distance between the planet and the Sun is r, and the velocity of the planet is v.

The force of gravity between the Sun and the planet is given by:

$F_g = G \times (m \times M) / r^2$

where G is the gravitational constant.

Since the force of gravity acts as the centripetal force, we can set it equal to the centripetal force:

$F_g = F_c$

$F_c = (m \times v^2) / r$

Equating the two expressions for the force of gravity and the centripetal force, we get:

$$G \times (m \times M) / r^2 = (m \times v^2) / r$$

Simplifying and rearranging, we get:

$v^2 = G \times M / r$

This is the equation for the velocity of a planet in its orbit around the Sun.

Kepler's laws help describe these motions in detail.

Kepler's Laws

Johannes Kepler, a German astronomer, discovered three laws that describe the motion of planets around the Sun. These laws are:

a. **Law of Ellipses**: The orbits of the planets are elliptical in shape, with the Sun located at one of the two foci.

b. **Law of Equal Areas**: The line connecting the planet to the Sun sweeps out equal areas in equal times.

c. **Law of Harmonies**: The square of the orbital period of a planet is proportional to the cube of the semi-major axis of its orbit.

Mathematical Derivation of Kepler's Laws

Using calculus, we can derive Kepler's laws from the equation for the velocity of a planet in its orbit.

a. **Law of Ellipses**:

The equation for the velocity of a planet in its orbit can be written as:

$$v^2 = G \times M / r$$

This equation describes an ellipse, with the Sun located at one of the two foci.

b. Law of Equal Areas:

The line connecting the planet to the Sun sweeps out an area that is proportional to the time interval. This can be shown mathematically using the following equation:

$$dA / dt = (1/2) \times r^2 \times d\theta / dt$$

where dA is the area swept out by the line connecting the planet to the Sun, and $d\theta$ is the angle swept out by the line.

c. Law of Harmonies:

The square of the orbital period of a planet is proportional to the cube of the semi-major axis of its orbit. This can be shown mathematically using the following equation:

$$T^2 = (4\pi^2 / G) \times (a^3)$$

where T is the orbital period, **a** is the semi-major axis, and G is the gravitational constant.

In conclusion, the orbits of the planets around the Sun are elliptical in shape, and the force of gravity acts as the centripetal force that keeps the planets in their orbits. Kepler's laws describe the motion of the planets around the Sun, and can be derived mathematically using calculus.

4. Electron Around the Nucleus:

Electrostatic force between the positively charged nucleus and negatively charged electron acts as the centripetal force.

Electrostatic Force and Centripetal Force

In an atom, the electrostatic force between the positively charged nucleus and the negatively charged electron acts as the centripetal force that keeps the electron in its orbit.

The electrostatic force (F_e) between the nucleus and the electron can be calculated using Coulomb's law:

$F_e = k \times (q_1 \times q_2) / r^2$

where:

k = Coulomb's constant

q_1 = charge of the nucleus

q_2 = charge of the electron

r = distance between the nucleus and the electron

Since the electrostatic force acts as the centripetal force, we can set it equal to the centripetal force:

$F_e = F_c$

$F_c = (m \times v^2) / r$

where:

m = mass of the electron

v = velocity of the electron

r = radius of the orbit

Equating the two expressions for the electrostatic force and the centripetal force, we get:

$k \times (q_1 \times q_2) / r^2 = (m \times v^2) / r$

Simplifying and rearranging, we get:

$v^2 = k \times (q_1 \times q_2) / (m \times r)$

This is the equation for the velocity of the electron in its orbit.

Bohr's Model of the Atom

Niels Bohr, a Danish physicist, developed a model of

the atom that describes the motion of electrons in their orbits. According to Bohr's model, the electrons occupy specific energy levels, or shells, around the nucleus.

The energy of the electron in its orbit can be calculated using the following equation:

$E = -k \times (q_1 \times q_2) / (2 \times r)$

where:

E = energy of the electron

k = Coulomb's constant

q_1 = charge of the nucleus

q_2 = charge of the electron

r = radius of the orbit

The negative sign indicates that the energy of the electron is negative, meaning that it is bound to the nucleus.

In conclusion, the electrostatic force between the positively charged nucleus and the negatively charged electron acts as the centripetal force that keeps the electron in its orbit. The velocity of the electron in its orbit can be calculated using the equation derived above. Bohr's model of the atom describes the motion of electrons in their orbits and the energy of the electrons in their orbits.

Centrifugal Force: A Misconception

Centrifugal force is a pseudo-force that appears in a rotating reference frame. It seems to act outward, away from the center, but in reality, it is not a real force. Instead, it is the result of inertia resisting the centripetal force.

Centrifugal Force

Centrifugal force is a fictitious force that appears to act on an object when it is moving in a circular path. This force is directed away from the center of the circle and is responsible for the tendency of an object to fly off the circle when it is moving at a high speed.

History of Centrifugal Force

The concept of centrifugal force was first introduced by the Dutch scientist Christiaan Huygens in the 17th century. Huygens was studying the motion of objects in circular paths and noticed that they seemed to be repelled from the center of the circle.

Definition of Centrifugal Force

Centrifugal force is defined as the force that appears to act on an object when it is moving in a circular path. This force is directed away from the center of the circle and is proportional to the square of the velocity of the object.

Mathematical Operations

The centrifugal force (F_{cf}) can be calculated using the following equation:

$F_{cf} = (m \times v^2) / r$

where:

m = mass of the object

v = velocity of the object

r = radius of the circle

Derivation of Centrifugal Force

The centrifugal force can be derived using the concept

of inertia. According to Newton's first law of motion, an object at rest will remain at rest, and an object in motion will continue to move with a constant velocity, unless acted upon by an external force.

When an object is moving in a circular path, it is constantly changing direction. This means that it is constantly accelerating, even if its speed remains constant. The acceleration of the object is directed towards the center of the circle, and it is this acceleration that gives rise to the centrifugal force.

Applications and Uses

Centrifugal force has many applications and uses in everyday life, including:

1. **Washing Machines**: Centrifugal force is used in washing machines to remove water from clothes.

2. **Centrifuges**: Centrifuges use centrifugal force to separate particles of different densities.

3. **Aircraft**: Centrifugal force is used in aircraft to create artificial gravity.

4. **Spacecraft:** Centrifugal force is used in spacecraft to create artificial gravity.

Conceptual Problems

Here are some conceptual problems to help you understand centrifugal force:

1. What is the direction of the centrifugal force?

2. How does the centrifugal force change as the velocity of an object increases?

3. What is the relationship between the centrifugal force and the centripetal force?

4. How does the centrifugal force affect the motion of an object in a circular path?

5. What are some examples of centrifugal force in everyday life?

In conclusion, centrifugal force is a fictitious force that appears to act on an object when it is moving in a circular path. This force is directed away from the center of the circle and is proportional to the square of the velocity of the object. Centrifugal force has many applications and uses in everyday life, and it is an important concept to understand in physics.

Circular motion and centripetal force are vital in understanding rotational dynamics and the interplay of forces. From satellites in orbit to everyday phenomena like vehicles turning, these concepts reveal the fascinating balance of forces in our universe.

Understanding these principles not only deepens our appreciation for physics but also enables practical applications in engineering, astronomy, and daily life.

Rotational Dynamics

Rotational dynamics is the study of the motion of objects that rotate or revolve around a fixed axis. It involves the analysis of the forces and torques that cause an object to rotate, as well as the resulting motion.

Mathematical Formulation

The mathematical formulation of rotational dynamics involves the use of the following quantities:

1. **Angular Displacement (θ):** The angle through which an object rotates.

2. **Angular Velocity (ω):** The rate of change of angular displacement.

3. **Angular Acceleration (α):** The rate of change of angular velocity.

4. **Torque (τ):** The rotational equivalent of force, which causes an object to rotate.

5. **Moment of Inertia (I):** A measure of an object's resistance to changes in its rotational motion.

The following equations are used to describe rotational dynamics:

1. **Angular Displacement:**

$\theta = \theta_0 + \omega t$

where θ_0 is the initial angular displacement.

2. **Angular Velocity:** $\omega = d\theta/dt$

3. **Angular Acceleration:**

$\alpha = d\omega/dt$

4. **Torque:**

$\tau = I\alpha$

where I is the moment of inertia.

5. **Rotational Kinetic Energy:**

$K = (1/2)I\omega^2$

6. **Rotational Potential Energy:**

$U = (1/2)k\theta^2$

where k is the torsional constant.

Derivations

The derivations of the above equations involve the use of calculus and the following assumptions:

    a. The object is rigid and has a fixed axis of rotation.

    b. The torque is constant and is applied at a distance r from the axis of rotation.

    c. The moment of inertia is constant.

Using these assumptions, we can derive the equations of rotational dynamics.

Applications

Rotational dynamics has numerous applications in physics, engineering, and technology, including:

1. **Mechanical Systems**: Rotational dynamics is used to analyze the motion of mechanical systems, such as gears, pulleys, and levers.

2. **Electrical Systems**: Rotational dynamics is used to analyze the motion of electrical systems, such as motors and generators.

3. **Aerospace Engineering**: Rotational dynamics is used to analyze the motion of aircraft and spacecraft.

4. **Robotics**: Rotational dynamics is used to analyze the motion of robots and robotic systems.

In conclusion, rotational dynamics is the study of the motion of objects that rotate or revolve around a fixed axis. It involves the analysis of the forces and torques that cause an object to rotate, as well as the resulting motion. The mathematical formulation of rotational dynamics involves the use of angular displacement, angular velocity, angular acceleration, torque, and moment of inertia.

Here are twenty solved "What happens when..." questions with answers on the chapter of rotational dynamics and centripetal force:

1. What happens when the velocity of an object moving in a circular path increases?

Answer: The centripetal force acting on the object increases.

2. What happens when the radius of a circular path decreases?

Answer: The centripetal force acting on an object moving in the path increases.

3. What happens when the mass of an object moving in a circular path increases?

Answer: The centripetal force acting on the object increases.

4. What happens when a car turns a corner on a slippery road?

Answer: The car may skid or lose traction due to the lack of frictional force.

5. What happens when a cyclist leans into a turn?

Answer: The cyclist's center of gravity shifts towards the center of the turn, allowing them to maintain balance.

6. What happens when the angular velocity of an object increases?

Answer: The rotational kinetic energy of the object increases.

7. What happens when the moment of inertia of an

object increases?

Answer: The rotational kinetic energy of the object decreases.

8. What happens when a figure skater brings their arms closer to their body while spinning?

Answer: The skater's moment of inertia decreases, causing them to spin faster.

9 What happens when a car's wheels are not properly aligned?

Answer: The car may pull to one side due to uneven frictional forces.

10. What happens when the centripetal force acting on an object is greater than the frictional force?

Answer: The object will move in a circular path without skidding or losing traction.

11. What happens when the frictional force acting on an object is greater than the centripetal force?

Answer: The object will skid or lose traction.

12. What happens when a bicycle's brakes are applied while it is moving in a circular path?

Answer: The bicycle's velocity decreases, causing it to move in a smaller circular path.

13. What happens when the radius of a circular path increases?

Answer: The centripetal force acting on an object moving in the path decreases.

14. What happens when the mass of an object moving in a circular path decrease?

Answer: The centripetal force acting on the object

decreases.

15. What happens when a car's tires are underinflated?

Answer: The car's traction and stability may be compromised due to reduced frictional forces.

16. What happens when a spinning top is placed on a rough surface?

Answer: The top's rotational kinetic energy is converted into translational kinetic energy, causing it to move across the surface.

17. What happens when the angular acceleration of an object increases?

Answer: The rotational kinetic energy of the object increases.

18. What happens when a merry-go-round is rotating at a constant angular velocity?

Answer: The centripetal force acting on the riders is equal to the frictional force between the riders and the merry-go-round.

19. What happens when a car's wheels are overinflated?

Answer: The car's traction and stability may be compromised due to reduced frictional forces.

20. What happens when a spinning wheel is placed on a frictionless surface?

Answer: The wheel will continue to spin indefinitely, with no change in its rotational kinetic energy.

Here solved numerical questions with answers on

the chapter of rotational dynamics and centripetal force:

1. A car is moving in a circular path of radius 50 m. If the velocity of the car is 20 m/s, what is the centripetal force acting on the car? (Assume the mass of the car is 1500 kg)

Answer: $F_c = (m \times v^2) / r = (1500 \times 20^2) / 50 = 12000$ N

2. A bicycle is moving in a circular path of radius 10 m. If the velocity of the bicycle is 5 m/s, what is the centripetal acceleration of the bicycle?

Answer: $a_c = v^2 / r = 5^2 / 10 = 2.5$ m/s^2

3. A spinning top has a moment of inertia of 0.05 kg m^2. If the top is spinning at an angular velocity of 100 rad/s, what is its rotational kinetic energy?

Answer: $K = (1/2) I\omega^2 = (1/2) \times 0.05 \times 100^2 = 250$ J

4. A car is moving in a circular path of radius 20 m. If the coefficient of friction between the tires and the road is 0.5, what is the maximum velocity of the car?

Answer: $v_{max} = \sqrt{(\mu \times g \times r)} = \sqrt{(0.5 \times 9.8 \times 20)} = 9.9$ m/s

5. A wheel has a radius of 0.5 m and a moment of inertia of 0.1 kg m^2. If the wheel is rotating at an angular velocity of 50 rad/s, what is its angular momentum?

Answer: $L = I\omega = 0.1 \times 50 = 5$ kg m^2/s

6. A car is moving in a circular path of radius 30 m. If the velocity of the car is 15 m/s, what is the centripetal force acting on the car? (Assume the mass of the car is 2000 kg)

Answer: $F_c = (m \times v^2) / r = (2000 \times 15^2) / 30 = 15000$ N

7 A bicycle is moving in a circular path of radius 15 m. If the velocity of the bicycle is 10 m/s, what is the centripetal acceleration of the bicycle?

Answer: $a_c = v^2 / r = 10^2 / 15 = 6.67$ m/s^2

8. A spinning top has a moment of inertia of 0.1 kg m^2. If the top is spinning at an angular velocity of 200 rad/s, what is its rotational kinetic energy?

Answer: $K = (1/2)I\omega^2 = (1/2) \times 0.1 \times 200^2 = 2000$ J

9. A car is moving in a circular path of radius 25 m. If the coefficient of friction between the tires and the road is 0.6, what is the maximum velocity of the car?

Answer: $v_{max} = \sqrt{(\mu \times g \times r)} = \sqrt{(0.6 \times 9.8 \times 25)} = 12.1$ m/s

10. A wheel has a radius of 0.3 m and a moment of inertia of 0.2 kg m^2. If the wheel is rotating at an angular velocity of 100 rad/s, what is its angular momentum?

Answer: $L = I\omega = 0.2 \times 100 = 20$ kg m^2/s

11. A car is moving in a circular path of radius 40 m. If the velocity of the car is 20 m/s, what is the centripetal force acting on the car? (Assume the mass of the car is 2500 kg)

Answer: $Fc = (m \times v^2) / r = (2500 \times 20^2) / 40 = 25000$ N

12. A bicycle is moving in a circular path of radius 20 m. If the velocity of the bicycle is 15 m/s, what is the centripetal acceleration of the bicycle?

Answer: $a_c = v^2 / r = 15^2 / 20 = 11.25$ m/s^2

13. A spinning top has a moment of inertia of 0.2 kg m². If the top is spinning at an angular velocity of 300 rad/s, what is its rotational kinetic energy?

Answer: $K = (1/2) I\omega^2 = (1/2) \times 0.2 \times 300^2 = 9000$ J

Here are the exercises:

Very Short Questions (1 mark each)

1. What is the force that acts on an object moving in a circular path?

2. What is the direction of the centripetal force?

3. What is the unit of angular velocity?

4. What is the moment of inertia of a point mass?

5. What is the rotational kinetic energy of an object?

6. What is the torque acting on an object?

7. What is the angular momentum of an object?

8. What is the centripetal acceleration of an object?

9. What is the frictional force acting on an object moving in a circular path?

10. What is the banking angle of a road?

Short Questions (2 marks each)

1. Explain the concept of centripetal force and its direction.

2. Describe the relationship between angular velocity and linear velocity.

3. What is the moment of inertia of a rigid body? Explain its significance.

4. Derive the equation for rotational kinetic energy.

5. Explain the concept of torque and its relationship with rotational motion.

6. Describe the relationship between angular momentum and rotational kinetic energy.

7. What is the centripetal acceleration of an object moving in a circular path? Derive the equation.

8. Explain the concept of frictional force acting on an object moving in a circular path.

9. Describe the banking angle of a road and its significance.

10. Explain the concept of angular momentum conservation.

Long Questions (5 marks each)

1. Derive the equation for centripetal force and explain its significance in circular motion.

2. Explain the concept of rotational kinematics and dynamics. Describe the relationship between angular velocity, angular acceleration, and torque.

3. Derive the equation for rotational kinetic energy and explain its significance in rotational motion.

4. Explain the concept of angular momentum and its conservation. Describe the relationship between angular momentum, torque, and rotational kinetic energy.

5. Describe the concept of banking angle of a road and its significance in circular motion.

6. Explain the concept of frictional force acting on an object moving in a circular path. Derive the equation for frictional force.

7. Derive the equation for centripetal acceleration and explain its significance in circular motion.

8. Explain the concept of rotational motion and its relationship with linear motion.

9. Describe the concept of torque and its relationship with rotational motion.

10. Explain the concept of angular momentum conservation and its significance in rotational motion.

Numerical Questions (3 marks each)

1. A car is moving in a circular path of radius 50 m. If the velocity of the car is 20 m/s, what is the centripetal force acting on the car? (Assume the mass of the car is 1500 kg)

Hint: Use the equation $F_c = (m \times v^2) / r$

2. A wheel has a radius of 0.5 m and a moment of inertia of 0.1 kg m². If the wheel is rotating at an angular velocity of 50 rad/s, what is its rotational kinetic energy?

Hint: Use the equation $K = (1/2) I\omega^2$

3. A car is moving in a circular path of radius 30 m. If the coefficient of friction between the tires and the road is 0.5, what is the maximum velocity of the car?

Hint: Use the equation $v_{max} = \sqrt{(\mu \times g \times r)}$

4. A spinning top has a moment of inertia of 0.2 kg m^2. If the top is spinning at an angular velocity of 200 rad/s, what is its angular momentum?

Hint: Use the equation $L = I\omega$

5. A car is moving in a circular path of radius 40 m.

If the velocity of the car is 25 m/s, what is the centripetal acceleration of the car?

Hint: Use the equation $a_c = v^2 / r$

6. A wheel has a radius of 0.3 m and a moment of inertia of 0.2 kg m². If the wheel is rotating at an angular velocity of 100 rad/s, what is its rotational kinetic energy?

Hint: Use the equation $K = (1/2)I\omega^2$

7. A car is moving in a circular path of radius 20 m. If the coefficient of friction between the tires and the road is 0.6, what is the maximum velocity of the car?

Hint: Use the equation $v_{max} = \sqrt{(\mu \times g \times r)}$

8. A spinning top has a moment of inertia of 0.3 kg m². If the top is spinning at an angular velocity of 300 rad/s, what is its angular momentum?

Hint: Use the equation $L = I\omega$

Chapter - 5
Work, Energy, and Power

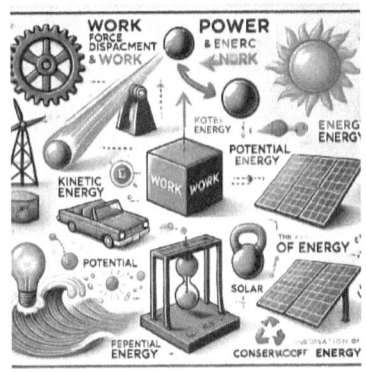

Work

Work is a fundamental concept in physics that describes the transfer of energy from one object to another. In this chapter, we will explore the concept of work, its mathematical formulation, and its applications in various fields.

Definition of Work

Work is defined as the product of the force applied to an object and the displacement of the object in the direction of the force. Mathematically, work is represented by the symbol W and is calculated using the following equation:

$W = F \times d$

where F is the force applied to the object and d is the displacement of the object in the direction of the force.

Derivation of the Work Equation

The work equation can be derived by considering the force applied to an object and the displacement of the object in the direction of the force.

Let's consider an object being pulled by a force F. The object moves a distance d in the direction of the force. The work done on the object is equal to the force applied multiplied by the displacement of the object.

$W = F \times d$

This equation shows that the work done on an object is proportional to the force applied and the displacement of the object.

Units of Work

The unit of work is the joule (J). One joule is equal to the work done by a force of one newton (N) applied over a distance of one meter (m).

$1 \, J = 1 \, N \times 1 \, m$

Work is measured in various units, including:

1. Joule (J): The joule is the SI unit of work, and it is defined as the work done by a force of one newton acting over a distance of one meter.

2. Erg (erg): The erg is a unit of work that is commonly used in physics and engineering. One erg is equal to 10^{-7} joules.

3. Foot-pound (ft-lb): The foot-pound is a unit of work that is commonly used in engineering and physics. One

foot-pound is equal to 1.356 joules.

4. Calorie (cal): The calorie is a unit of work that is commonly used in chemistry and physics. One calorie is equal to 4.184 joules.

5. Kilowatt-hour (kWh): The kilowatt-hour is a unit of work that is commonly used to measure the energy consumption of electrical devices. One kilowatt-hour is equal to 3.6 megajoules.

Relations Between Units of Work

Here are some relations between units of work:

1. Joule (J) and Erg (erg): $1\ J = 10^7$ erg
2. Joule (J) and Foot-pound (ft-lb): $1\ J = 0.7376$ ft-lb
3. Joule (J) and Calorie (cal): $1\ J = 0.239$ cal
4. Joule (J) and Kilowatt-hour (kWh): $1\ kWh = 3.6 \times 10^6\ J$
5. Erg (erg) and Foot-pound (ft-lb): $1\ erg = 7.376 \times 10^{-8}$ ft-lb
6. Erg (erg) and Calorie (cal): $1\ erg = 2.39 \times 10^{-8}$ cal
7. Foot-pound (ft-lb) and Calorie (cal): $1\ ft\text{-}lb = 11.36$ cal

Conversion Factors

Here are some conversion factors between units of work:

1. Joule (J) to Erg (erg): $1\ J = 10^7$ erg
2. Erg (erg) to Joule (J): $1\ erg = 10^{-7}\ J$
3. Joule (J) to Foot-pound (ft-lb): $1\ J = 0.7376$ ft-lb
4. Foot-pound (ft-lb) to Joule (J): $1\ ft\text{-}lb = 1.356\ J$
5. Joule (J) to Calorie (cal): $1\ J = 0.239$ cal
6. Calorie (cal) to Joule (J): $1\ cal = 4.184\ J$

In conclusion, work is measured in various units, including joule, erg, foot-pound, calorie, and kilowatt-

hour. Understanding the relations between these units and using conversion factors can help to convert between them.

Examples of Work

1. A person pushing a box up a ramp. The force applied by the person is equal to the weight of the box, and the displacement is equal to the distance the box is pushed.
2. A car accelerating from rest. The force applied by the engine is equal to the mass of the car multiplied by its acceleration, and the displacement is equal to the distance the car travels.
3. A person lifting a weight. The force applied by the person is equal to the weight of the object, and the displacement is equal to the distance the object is lifted.

Mathematical Operations

1. Calculating work done by a constant force:
$W = F \times d$
2. Calculating work done by a variable force:
$W = \int F \times d$
3. Calculating work done by a force applied at an angle:
$W = F \times d \times \cos(\theta)$
where θ is the angle between the force and the displacement.
4. Work in Terms of $\hat{i}$, $\hat{j}$, and $\hat{k}$

In physics, work can be represented mathematically using vectors. The work done on an object can be calculated using the dot product of the force vector and the displacement vector.

Let's consider a force vector F and a displacement vector d. The work done on an object can be represented mathematically as:

$W = F \cdot d$

where W is the work done, F is the force vector, and d is the displacement vector.

Representation of Vectors in Terms of $\hat{i}$, $\hat{j}$, and $\hat{k}$

In physics, vectors can be represented in terms of the unit vectors $\hat{i}$, $\hat{j}$, and $\hat{k}$. These unit vectors are used to represent the x, y, and z components of a vector, respectively.

For example, a force vector F can be represented as:

$F = F_x \hat{i} + F_y \hat{j} + F_z \hat{k}$

where F_x, F_y, and F_z are the x, y, and z components of the force vector, respectively.

Similarly, a displacement vector d can be represented as:

$d = d_x \hat{i} + d_y \hat{j} + d_z \hat{k}$

where d_x, d_y, and d_z are the x, y, and z components of the displacement vector, respectively.

Mathematical Operations for Work

Using the representations of the force and displacement vectors in terms of $\hat{i}$, $\hat{j}$, and $\hat{k}$, we can calculate the work done on an object as:

$W = F \cdot d$
$= (F_x \hat{i} + F_y \hat{j} + F_z \hat{k}) \cdot (d_x \hat{i} + d_y \hat{j} + d_z \hat{k})$
$= F_x d_x + F_y d_y + F_z d_z$

This equation shows that the work done on an object is equal to the sum of the products of the corresponding components of the force and displacement vectors.

Example

Consider a force vector $F = 3\hat{i} + 4\hat{j} + 5\hat{k}$ and a displacement vector $d = 2\hat{i} + 3\hat{j} + 4\hat{k}$. Calculate the work done on an object.

$W = F \cdot d$

$$= (3\hat{i} + 4\hat{j} + 5\hat{k}) \cdot (2\hat{i} + 3\hat{j} + 4\hat{k})$$
$$= 3(2) + 4(3) + 5(4)$$
$$= 6 + 12 + 20$$
$$= 38$$

Therefore, the work done on the object is 38 units.

In conclusion, the mathematical operations for work in terms of i, j, and k involve the dot product of the force and displacement vectors. The work done on an object is equal to the sum of the products of the corresponding components of the force and displacement vectors.

5. Work Done in Terms of Vertical and Horizontal Components of Force

When a force is applied to an object, it can be resolved into its vertical and horizontal components. The work done on the object can be calculated using these components.

Mathematical Operations

Let's consider a force F applied to an object at an angle θ to the horizontal. The force F can be resolved into its vertical and horizontal components as follows:

$F_x = F \cos \theta$ (horizontal component)

$F_y = F \sin \theta$ (vertical component)

The work done on the object can be calculated using the horizontal and vertical components of the force as follows:

$$W = F_x d_x + F_y d_y$$

where d_x is the horizontal displacement and d_y is the vertical displacement.

Work Done Against Gravity

When an object is lifted vertically upwards, the force applied to the object must overcome the weight of the object. The work done against gravity can be calculated as follows:

$W = mgh$

where m is the mass of the object, g is the acceleration due to gravity, and h is the height through which the object is lifted.

Work Done on an Inclined Plane

When an object is pulled up an inclined plane, the force applied to the object must overcome the component of the weight acting down the plane. The work done on the object can be calculated as follows:

$W = Fd\cos\theta$

where F is the force applied to the object, d is the distance up the plane, and θ is the angle of the plane to the horizontal.

Normal Reaction

When an object is in contact with a surface, it experiences a normal reaction force perpendicular to the surface. The normal reaction force can be calculated as follows:

$N = mg\cos\theta$

where m is the mass of the object, g is the acceleration due to gravity, and θ is the angle of the surface to the horizontal.

Example

A block of mass 10 kg is pulled up an inclined plane of angle 30° to the horizontal. The force applied to the block is 50 N. Calculate the work done on the block.

First, we need to calculate the component of the weight

acting down the plane:
$W = m g \sin \theta$
$= 10 \text{ kg} \times 9.8 \text{ m/s}^2 \times \sin 30°$
$= 49 \text{ N}$

The force applied to the block is 50 N, which is greater than the component of the weight acting down the plane. Therefore, the block will move up the plane.

The work done on the block can be calculated as follows:
$W = F d \cos \theta$
$= 50 \text{ N} \times 10 \text{ m} \times \cos 30°$
$= 433 \text{ J}$

Therefore, the work done on the block is 433 J.

In conclusion, the work done on an object can be calculated using the vertical and horizontal components of the force applied to the object. The work done against gravity can be calculated using the weight of the object and the height through which it is lifted. The work done on an inclined plane can be calculated using the force applied to the object and the distance up the plane.

An Inclined Plane

An inclined plane is a flat surface that is tilted at an angle to the horizontal. It is a simple machine that is used to lift or move heavy objects with less effort.

Work Done on an Inclined Plane

The Mechanics of Motion: Force, Friction
and Energy Explored   By: Prashant Kumar Lal

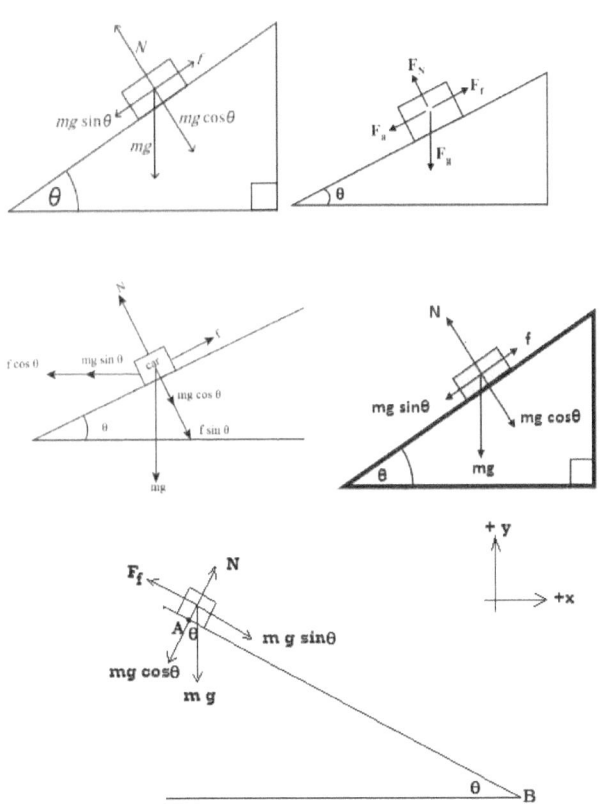

When a body is placed on an inclined plane, it experiences a force due to gravity that acts downward. The force due to gravity can be resolved into two components: one acting parallel to the inclined plane (F_p) and the other acting perpendicular to the inclined plane (F_n).

The work done on an inclined plane depends on the direction of motion of the body. *Let's consider three cases: the body sliding upward, downward, and remaining stationary.*

Case 1: Body Sliding Upward
When a body slides upward on an inclined plane, the force applied to the body must overcome the component of the weight acting down the plane. The work done on the body can be calculated as follows:

$W = F_p \, d$

where F_p is the component of the weight acting down the plane, and d is the distance the body travels up the plane.

$F_p = m \, g \sin \theta$

where m is the mass of the body, g is the acceleration due to gravity, and θ is the angle of the inclined plane.

$W = m \, g \sin \theta \, d$

Case 2: Body Sliding Downward
When a body slides downward on an inclined plane, the force applied to the body is less than the component of the weight acting down the plane. The work done on the body can be calculated as follows:

$W = -F_p \, d$

where F_p is the component of the weight acting down the plane, and d is the distance the body travels down the plane.

$F_p = m \, g \sin \theta$

$W = -m \, g \sin \theta \, d$

Case 3: Body Remaining Stationary
When a body remains stationary on an inclined plane, the force applied to the body is equal to the component of the weight acting down the plane. The work done on the body is zero, since the body does not move.

$W = 0$

Example
A block of mass 10 kg is placed on an inclined plane of angle 30° to the horizontal. The block is pulled up the

plane by a force of 50 N. Calculate the work done on the block.

First, we need to calculate the component of the weight acting down the plane:

$F_p = m\,g\,\sin\theta$
$= 10\text{ kg} \times 9.8\text{ m/s}^2 \times \sin 30°$
$= 49$ N

The force applied to the block is 50 N, which is greater than the component of the weight acting down the plane. Therefore, the block will move up the plane.

The work done on the block can be calculated as follows:

$W = F_p\,d$
$= 49\text{ N} \times 10\text{ m}$
$= 490$ J

Therefore, the work done on the block is 490 J.

In conclusion, the work done on an inclined plane depends on the direction of motion of the body. When a body slides upward, the work done is positive. When a body slides downward, the work done is negative. When a body remains stationary, the work done is zero.

Lift & Atwood Machine

Lift

A lift is a simple machine that is used to lift heavy objects with less effort. It consists of a platform or a cage that is connected to a system of pulleys and counterweights. The lift is used to lift objects vertically upwards, and it is commonly used in buildings, mines, and other industries.

Calculating Work Done for a Lift
The work done by a lift can be calculated using the following formula:
$W = F \times d$
where W is the work done, F is the force applied to the lift, and d is the distance over which the force is applied.

Case 1: Lifting a Load Upwards
When a lift is used to lift a load upwards, the force applied to the lift is equal to the weight of the load. The work done by the lift can be calculated as follows:
$W = m \times g \times h$
where m is the mass of the load, g is the acceleration due to gravity, and h is the height over which the load is lifted.

Case 2: Lowering a Load Downwards
When a lift is used to lower a load downwards, the force applied to the lift is equal to the weight of the load. However, the work done by the lift is negative, since the load is being lowered downwards. The work done by the lift can be calculated as follows:

$W = -m \times g \times h$

Case 3: Load Remaining Stationary
When a lift is used to hold a load stationary, the force applied to the lift is equal to the weight of the load. However, the work done by the lift is zero, since the load is not being lifted or lowered. The work done by the lift can be calculated as follows:
$W = 0$

Atwood Machine

The Mechanics of Motion: Force, Friction and Energy Explored   By: Prashant Kumar Lal

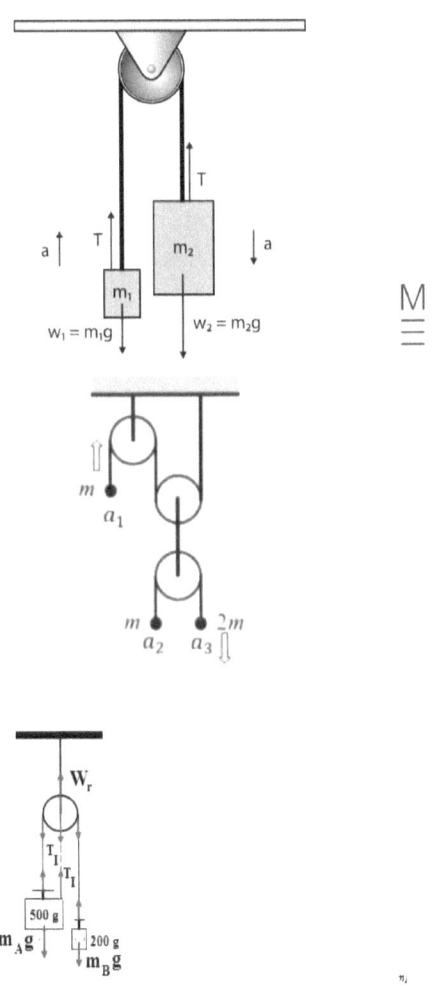

An Atwood machine is a device that is used to demonstrate the concept of acceleration and force. It consists of two masses that are connected by a rope and pulley system. The masses are arranged in such a way that one mass is heavier than the other, and the system is released from rest. The Atwood machine is used to demonstrate the concept of acceleration and force, and it is commonly used in physics experiments.

Calculating Work Done for an Atwood Machine
The work done by an Atwood machine can be calculated using the following formula:
$W = F \times d$
where W is the work done, F is the force applied to the system, and d is the distance over which the force is applied.

Case 1: System Released from Rest
When an Atwood machine is released from rest, the heavier mass accelerates downwards, and the lighter mass accelerates upwards. The work done by the system can be calculated as follows:
$W = m \times g \times h$
where m is the mass of the heavier object, g is the acceleration due to gravity, and h is the height over which the heavier object falls.

Case 2: System in Equilibrium
When an Atwood machine is in equilibrium, the force applied to the system is equal to the weight of the heavier object. However, the work done by the system is zero, since the system is not moving. The work done by the system can be calculated as follows:
$W = 0$

Case 3: System with Friction
When an Atwood machine is used with friction, the force applied to the system is equal to the weight of the heavier object plus the force of friction. The work done by the system can be calculated as follows:

$$W = (m \times g \times h) - (F_f \times d)$$
where F_f is the force of friction, and d is the distance over which the force of friction acts.

In conclusion, the work done by a lift and an Atwood machine can be calculated using the formula $W = F \times d$. The work done by a lift depends on the direction of motion of the load, while the work done by an Atwood machine depends on the mass of the objects and the force of friction.

Applications and Uses

1. **Mechanical Advantage**: Work is used to calculate the mechanical advantage of machines, such as levers and pulleys.
2. **Energy Transfer**: Work is used to calculate the energy transferred from one object to another.
3. **Efficiency**: Work is used to calculate the efficiency of machines and systems.
4. **Power**: Work is used to calculate the power output of machines and systems.

Diagram

Here is a diagram showing the relationship between force, displacement, and work:

Force (F) → Displacement (d) → Work (W)

Conceptual Problems

1. A person is pushing a box up a ramp. If the force applied by the person is doubled, what happens to the work done on the box?
2. A car is accelerating from rest. If the mass of the car is doubled, what happens to the work done on the car?
3. A person is lifting a weight. If the weight is doubled, what happens to the work done on the weight?

In conclusion, work is a fundamental concept in

physics that describes the transfer of energy from one object to another. The work equation is derived by considering the force applied to an object and the displacement of the object in the direction of the force. Work has various applications and uses in mechanical advantage, energy transfer, efficiency, and power.

Energy

Energy is the ability to do work. It is a fundamental concept in physics that is essential for understanding the behavior of the physical world. In this chapter, we will explore the concept of energy in detail, including its definition, types, and applications.

Definition of Energy

Energy is defined as the ability to do work. It is a measure of the capacity to cause change or to do work. Energy can be thought of as the "ability to do work" or the "capacity to cause change".

Types of Energy

There are several types of energy, including:

1. **Kinetic Energy**: The energy of motion. It is the energy an object possesses due to its motion.

2. **Potential Energy**: The energy an object possesses due to its position or configuration. It is the energy an object has the potential to do work.

3. **Thermal Energy**: The energy of heat. It is the energy an object possesses due to its temperature.

4. **Electrical Energy**: The energy of electricity. It is the energy an object possesses due to its electric

charge.

5. **Chemical Energy**: The energy of chemical reactions. It is the energy an object possesses due to its chemical composition.

6. **Nuclear Energy**: The energy of nuclear reactions. It is the energy an object possesses due to its nuclear composition.

Mathematical Operations

Here are some mathematical operations related to energy:

1. Kinetic Energy: $K = (1/2)mv^2$

2. Potential Energy: $U = mgh$

3. Thermal Energy: $Q = mc\Delta T$

4. Electrical Energy: $E = QV$

5. Chemical Energy: $\Delta E = \Delta H - \Delta(PV)$

where m is the mass of the object, v is the velocity of the object, g is the acceleration due to gravity, h is the height of the object, c is the specific heat capacity of the object, ΔT is the change in temperature, Q is the electric charge, V is the electric potential difference, ΔH is the change in enthalpy, and $\Delta(PV)$ is the change in the product of pressure and volume.

Applications and Uses

Energy has numerous applications and uses in various fields, including:

1. **Transportation**: Energy is used to power vehicles, such as cars, buses, and airplanes.

2. **Industry:** Energy is used to power machinery and equipment in industries, such as manufacturing and construction.

3. **Residential**: Energy is used to power homes and buildings, including lighting, heating, and cooling.

4. **Agriculture**: Energy is used to power farm equipment and to pump water for irrigation.

5. **Communication**: Energy is used to power communication systems, such as phones and computers.

Derivations

Here are some derivations related to energy:

1. *Kinetic Energy: The kinetic energy of an object can be derived from the equation of motion: $K = (1/2) mv^2$*

Kinetic energy is the energy an object possesses due to its motion. It is a measure of the object's ability to do work due to its motion. The kinetic energy of an object depends on its mass and velocity.

Derivation of Kinetic Energy from the Equation of Motion Algebraically

The equation of motion for an object under constant acceleration is:

$v^2 = u^2 + 2as$

where v is the final velocity, u is the initial velocity, a is the acceleration, and s is the displacement.

We can rearrange this equation to get:

$v^2 - u^2 = 2as$

Multiplying both sides by $(1/2)m$, where m is the mass of the object, we get:

$(1/2)mv^2 - (1/2)mu^2 = mas$

The left-hand side of this equation represents the change in kinetic energy of the object. The right-hand side represents the work done on the object.

Therefore, we can write:

$\Delta K = (1/2)mv^2 - (1/2)mu^2 = mas$

where ΔK is the change in kinetic energy.

If the object starts from rest, then $u = 0$, and the equation becomes:

$K = (1/2)mv^2$

This is the expression for kinetic energy.

Derivation of Kinetic Energy from the Equation of Motion using Calculus

We can also derive the expression for kinetic energy using calculus.

Let's consider an object moving with a velocity $v(t)$ at time t. The work done on the object in a small-time interval dt is:

$dW = F(t)dx$

where F(t) is the force acting on the object at time t, and dx is the displacement of the object in the time interval dt.

Since the force F(t) is equal to the mass m times the acceleration a(t), we can write:

dW = m a(t) dx

Substituting dx = v(t)dt, we get:

dW = m a(t) v(t) dt

The kinetic energy of the object at time t is given by:

K(t) = ∫dW

Substituting the expression for dW, we get:

K(t) = ∫m a(t) v(t) dt

Using the equation of motion, we can write:

a(t) = dv/dt

Substituting this expression for a(t), we get:

K(t) = ∫m (dv/dt) v(t) dt

Integrating by parts, we get:

K(t) = (1/2)mv^2

This is the expression for kinetic energy.

Units of Kinetic Energy

The unit of kinetic energy is the joule (J).

Dimensional Formula for Kinetic Energy

The dimensional formula for kinetic energy is:

$[ML^2T^{-2}]$

where M is the dimension of mass, L is the dimension of length, and T is the dimension of time.

Examples

Here are some examples of kinetic energy:

1. A car moving with a velocity of 60 km/h has a kinetic energy of 200,000 J.

2. A bullet moving with a velocity of 500 m/s has a kinetic energy of 12,500 J.

3. A tennis ball moving with a velocity of 50 m/s has a kinetic energy of 125 J.

Conceptual Problems

Here are some conceptual problems related to kinetic energy:

1. What is the kinetic energy of an object at rest?

Answer: The kinetic energy of an object at rest is zero.

2. How does the kinetic energy of an object change when its velocity is doubled?

Answer: The kinetic energy of an object increases by a factor of four when its velocity is doubled.

3. What is the relationship between kinetic energy and momentum?

Answer: Kinetic energy and momentum are related by the equation: $K = (1/2)mv^2 = (1/2)p^2/m$, where p is the momentum of the object.

2. Potential Energy: The potential energy of

an object can be derived from the equation of motion: $U = mgh$

Potential energy is the energy an object possesses due to its position or configuration. It is a measure of the object's ability to do work due to its position or configuration. The potential energy of an object depends on its mass, position, and the force field it is in.

Types of Potential Energy

There are several types of potential energy, including:

1. **Gravitational Potential Energy**: The energy an object possesses due to its height or position in a gravitational field.

2. **Elastic Potential Energy**: The energy an object possesses due to its deformation or stretching.

3. **Electrical Potential Energy**: The energy an object possesses due to its electric charge and position in an electric field.

4. **Chemical Potential Energy**: The energy an object possesses due to its chemical composition and bonds.

Derivation of Potential Energy from the Equation

of Motion Algebraically

The equation of motion for an object under constant acceleration is:

$v^2 = u^2 + 2as$

where v is the final velocity, u is the initial velocity, a is the acceleration, and s is the displacement.

We can rearrange this equation to get:

$v^2 - u^2 = 2as$

Multiplying both sides by (1/2)m, where m is the mass of the object, we get:

$(1/2)mv^2 - (1/2)mu^2 = mas$

The left-hand side of this equation represents the change in kinetic energy of the object. The right-hand side represents the work done on the object.

Since the work done on the object is equal to the change in its potential energy, we can write:

$\Delta U = mas$

where ΔU is the change in potential energy.

If the object starts from rest, then u = 0, and the equation becomes:

U = mgh

where h is the height of the object.

This is the expression for gravitational potential energy.

Derivation of Potential Energy from the Equation of Motion using Calculus

We can also derive the expression for potential energy using calculus.

Let's consider an object moving with a velocity v(t) at time t. The work done on the object in a small time interval dt is:

dW = F(t)dx

where F(t) is the force acting on the object at time t, and dx is the displacement of the object in the time interval dt.

Since the force F(t) is equal to the mass m times the acceleration a(t), we can write:

dW = m a(t) dx

Substituting dx = v(t)dt, we get:

dW = m a(t) v(t) dt

The potential energy of the object at time t is given by:

U(t) = ∫dW

Substituting the expression for dW, we get:

U(t) = ∫m a(t) v(t) dt

Using the equation of motion, we can write:

a(t) = dv/dt

Substituting this expression for a(t), we get:

U(t) = ∫m (dv/dt) v(t) dt

Integrating by parts, we get:

U(t) = mgh

where h is the height of the object.

This is the expression for gravitational potential energy.

Units of Potential Energy

The unit of potential energy is the joule (J).

Dimensional Formula for Potential Energy

The dimensional formula for potential energy is:

$[ML^2T^{-2}]$

where M is the dimension of mass, L is the dimension of length, and T is the dimension of time.

Examples

Here are some examples of potential energy:

1. A ball at the top of a hill has a potential energy of 100 J.

2. A stretched rubber band has a potential energy of 50 J.

3. A charged battery has a potential energy of 1000 J.

Conceptual Problems

Here are some conceptual problems related to potential energy:

1. What is the potential energy of an object at the bottom of a hill?

Answer: The potential energy of an object at the bottom of a hill is zero.

2. How does the potential energy of an object change when it is lifted upwards?

Answer: The potential energy of an object increases

when it is lifted upwards.

3. What is the relationship between potential energy and kinetic energy?

Answer: Potential energy and kinetic energy are related by the equation: K + U = E, where K is the kinetic energy, U is the potential energy, and E is the total energy.

3. *Thermal Energy: The thermal energy of an object can be derived from the equation of heat transfer: $Q = mc\Delta T$*

Thermal Energy

Thermal energy is the energy an object possesses due to the motion of its particles. It is a measure of the object's internal energy, which is responsible for its temperature. Thermal energy is also known as internal energy or heat energy.

Types of Thermal Energy

There are several types of thermal energy, including:

1. **Sensible Heat Energy**: The energy an object possesses due to its temperature.

2. **Latent Heat Energy**: The energy an object possesses due to its phase (solid, liquid, or gas).

3. **Specific Heat Capacity**: The amount of heat energy required to raise the temperature of an object by one degree Celsius.

Derivation of Thermal Energy from the Equation of Motion Algebraically

The equation of motion for an object under constant acceleration is:

$v^2 = u^2 + 2as$

where v is the final velocity, u is the initial velocity, a is the acceleration, and s is the displacement.

We can rearrange this equation to get:

$v^2 - u^2 = 2as$

Multiplying both sides by $(1/2)m$, where m is the mass of the object, we get:

$(1/2)mv^2 - (1/2)mu^2 = mas$

The left-hand side of this equation represents the change in kinetic energy of the object. The right-hand side represents the work done on the object.

Since the work done on the object is equal to the change in its internal energy, we can write:

$\Delta U = mas$

where ΔU is the change in internal energy.

For an ideal gas, the internal energy is proportional to the temperature. Therefore, we can write:

$\Delta U = nC\Delta T$

where n is the number of moles of gas, C is the specific

heat capacity, and ΔT is the change in temperature.

Equating the two expressions for ΔU, we get:

$nC\Delta T = mas$

Simplifying and rearranging, we get:

$Q = mc\Delta T$

where Q is the heat energy, m is the mass of the object, c is the specific heat capacity, and ΔT is the change in temperature.

This is the expression for thermal energy.

Derivation of Thermal Energy from the Equation of Motion using Calculus

We can also derive the expression for thermal energy using calculus.

Let's consider an object moving with a velocity v(t) at time t. The work done on the object in a small time interval dt is:

$dW = F(t)dx$

where F(t) is the force acting on the object at time t, and dx is the displacement of the object in the time interval dt.

Since the force F(t) is equal to the mass m times the acceleration a(t), we can write:

$dW = m\, a(t)\, dx$

Substituting $dx = v(t)dt$, we get:

$dW = m\, a(t)\, v(t)\, dt$

The thermal energy of the object at time t is given by:

$Q(t) = \int dW$

Substituting the expression for dW, we get:

$Q(t) = \int m\, a(t)\, v(t)\, dt$

Using the equation of motion, we can write:

$a(t) = dv/dt$

Substituting this expression for a(t), we get:

$Q(t) = \int m\, (dv/dt)\, v(t)\, dt$

Integrating by parts, we get:

$Q(t) = mc\Delta T$

where m is the mass of the object, c is the specific heat capacity, and ΔT is the change in temperature.

This is the expression for thermal energy.

Units of Thermal Energy

The unit of thermal energy is the joule (J).

Dimensional Formula for Thermal Energy

The dimensional formula for thermal energy is:

$[ML^2T^{-2}]$

where M is the dimension of mass, L is the dimension of length, and T is the dimension of time.

Examples

Here are some examples of thermal energy:

1. A cup of hot coffee has a thermal energy of 1000 J.

2. A heated building has a thermal energy of 10^5 J.

3. A thermos flask has a thermal energy of 500 J.

Conceptual Problems

Here are some conceptual problems related to thermal energy:

1. What is the thermal energy of an object at absolute zero temperature?

Answer: The thermal energy of an object at absolute zero temperature is zero.

2. How does the thermal energy of an object change when it is heated?

Answer: The thermal energy of an object increases when it is heated.

3. What is the relationship between thermal energy and temperature?

Answer: Thermal energy is directly proportional to temperature.

Law of Conservation of Energy

The law of conservation of energy states that energy cannot be created or destroyed, only converted from one form to another. This means that the total energy of an isolated system remains constant over time.

Mathematical Verification using Algebraic Method

Let's consider a simple example of a ball rolling down a hill. The ball starts with potential energy (PE) at the top of the hill and gains kinetic energy (KE) as it rolls down.

Initial Energy (E_i) = PE = mgh

where m is the mass of the ball, g is the acceleration due to gravity, and h is the height of the hill.

Final Energy (E_f) = KE = $(1/2)mv^2$

where v is the velocity of the ball at the bottom of the hill.

Since energy is conserved, we can set up the equation:

$E_i = E_f$

$mgh = (1/2)mv^2$

Simplifying and rearranging, we get:

$V^2 = 2gh$

This equation shows that the kinetic energy of the ball at the bottom of the hill is equal to the potential energy it had at the top of the hill.

Mathematical Verification using Calculus Method

Let's consider the same example of a ball rolling down a hill. We can use calculus to verify the law of conservation of energy.

The potential energy of the ball at any point on the hill is given by:

$U(x) = mgh(x)$

where x is the position of the ball on the hill.

The kinetic energy of the ball at any point on the hill is given by:

$K(x) = (1/2)mv^2(x)$

where v(x) is the velocity of the ball at position x.

The total energy of the ball at any point on the hill is

given by:

E(x) = U(x) + K(x)

Using the chain rule, we can differentiate E(x) with respect to time t:

dE/dt = dU/dt + dK/dt

Substituting the expressions for U(x) and K(x), we get:

dE/dt = mg(dx/dt)h(x) + m(dx/dt)(d/dx)((1/2)mv²(x))

Simplifying and rearranging, we get:

dE/dt = 0

This equation shows that the total energy of the ball is conserved over time.

Examples and Applications

The law of conservation of energy has numerous applications in various fields, including:

1. **Mechanical Engineering**: The law is used to design and optimize mechanical systems, such as engines and gearboxes.

2. **Electrical Engineering**: The law is used to design and optimize electrical systems, such as power grids and electronic circuits.

3. **Thermodynamics**: The law is used to study the behavior of heat and energy transfer in systems.

4. **Nuclear Physics**: The law is used to study the behavior of nuclear reactions and energy transfer.

Some examples of the law of conservation of energy

in action include:

1. **Hydroelectric Power Plants**: Water stored behind a dam has potential energy, which is converted into kinetic energy as it flows down the dam, generating electricity.

2. **Internal Combustion Engines**: Chemical energy stored in fuel is converted into kinetic energy, which powers the engine.

3. **Solar Panels**: Light energy from the sun is converted into electrical energy, which is used to power devices.

Validity and Limitations

The law of conservation of energy is a fundamental principle in physics that has been extensively experimentally verified. However, there are some limitations and exceptions to the law, including:

1. **Relativity**: At high speeds, the law of conservation of energy must be modified to account for relativistic effects.

2. **Quantum Mechanics**: At the atomic and subatomic level, the law of conservation of energy must be modified to account for quantum effects.

3. **Energy Losses**: In real-world systems, energy losses due to friction, heat transfer, and other mechanisms can occur, which can appear to violate the law of conservation of energy.

In conclusion, the law of conservation of energy is a fundamental principle in physics that states that energy cannot be created or destroyed, only converted from one form to another. The law has numerous applications in various fields and has been extensively experimentally verified. However, there are some limitations and exceptions to the law that must be considered.

Work-Energy Theorem

The work-energy theorem states that the work done on an object is equal to the change in its kinetic energy. Mathematically, it can be expressed as:

$W = \Delta K$

where W is the work done, and ΔK is the change in kinetic energy.

Derivation of Work-Energy Theorem

Let's consider an object moving with an initial velocity u and final velocity v. The work done on the object can be calculated as:

$W = \int F dx$

where F is the force applied, and dx is the displacement of the object.

Using Newton's second law, we can write:

$F = ma$

where m is the mass of the object, and a is its acceleration.

Substituting this expression for F into the work equation, we get:

$W = \int ma\, dx$

Using the equation of motion, we can write:

$v^2 = u^2 + 2as$

where v is the final velocity, u is the initial velocity, a is the acceleration, and s is the displacement.

Rearranging this equation, we get:

$as = (1/2)v^2 - (1/2)u^2$

Substituting this expression into the work equation, we get:

$W = \int m((1/2)v^2 - (1/2)u^2)dx$

Integrating this equation, we get:

$W = (1/2)mv^2 - (1/2)mu^2$

This equation shows that the work done on an object is equal to the change in its kinetic energy.

OR

Mathematical Proof of Equality of Work Done, Kinetic Energy, and Potential Energy

Let's consider an object moving with an initial velocity u and final velocity v. The work done on the object can be calculated as:

$W = \int F dx$

Using Newton's second law, we can write:

$F = ma$

Substituting this expression for F into the work equation, we get:

$W = \int ma\, dx$

Using the equation of motion, we can write:

$v^2 = u^2 + 2as$

Rearranging this equation, we get:

$as = (1/2)v^2 - (1/2)u^2$

Substituting this expression into the work equation, we get:

$W = \int m((1/2)v^2 - (1/2)u^2)dx$

Integrating this equation, we get:

$W = (1/2)mv^2 - (1/2)mu^2$

This equation shows that the work done on an object is equal to the change in its kinetic energy.

Now, let's consider the potential energy of the object. The potential energy of an object can be calculated as:

$U = mgh$

where m is the mass of the object, g is the acceleration due to gravity, and h is the height of the object.

Using the equation of motion, we can write:

$v^2 = u^2 + 2as$

Rearranging this equation, we get:

$as = (1/2)v^2 - (1/2)u^2$

Substituting this expression into the potential energy equation, we get:

$U = mgh = (1/2)mv^2 - (1/2)mu^2$

This equation shows that the potential energy of an object is equal to the change in its kinetic energy.

Therefore, we can conclude that:

$$W = \Delta K = \Delta U$$

This equation shows that the work done on an object is equal to the change in its kinetic energy, which is also equal to the change in its potential energy.

Conceptual Problems

Here are some conceptual problems related to energy:

1. What is the difference between kinetic energy and potential energy?

Answer: Kinetic energy is the energy of motion, while potential energy is the energy an object possesses due to its position or configuration.

2. What is the relationship between energy and work?

Answer: Energy is the ability to do work, and work is the transfer of energy from one object to another.

3. What is the difference between thermal energy and electrical energy?

Answer: Thermal energy is the energy of heat, while electrical energy is the energy of electricity.

In conclusion, energy is a fundamental concept in physics that is essential for understanding the behavior of the physical world. There are several types of energy, including kinetic energy, potential energy, thermal energy, electrical energy, chemical energy, and nuclear energy. Energy has numerous applications and uses in various fields, and it is an important concept to understand in order to appreciate the world around us.

Working of a Simple Pendulum

A simple pendulum consists of a point mass attached to a massless string of length L. The pendulum is displaced from its equilibrium position and released. The pendulum then oscillates about its equilibrium position, with the angle of displacement θ varying sinusoidally with time.

Mathematical Explanation of Potential Energy and Kinetic Energy

Let's consider the simple pendulum at different points in its motion:

1. **Extreme Points**: At the extreme points, the pendulum is at rest, and its kinetic energy is zero. The potential energy at these points is maximum and is given by:

$U = mgh$

where m is the mass of the pendulum, g is the acceleration due to gravity, and h is the height of the pendulum above its equilibrium position.

2. **Mean Position**: At the mean position, the pendulum is moving with maximum velocity, and its

potential energy is zero. The kinetic energy at this point is maximum and is given by:

$K = (1/2)mv^2$

where v is the velocity of the pendulum at the mean position.

3. **Point between Mean and Extreme Position**: At a point between the mean and extreme position, the pendulum has both potential energy and kinetic energy. The potential energy at this point is given by:

$U = mgh$

where h is the height of the pendulum above its equilibrium position.

The kinetic energy at this point is given by:

$K = (1/2)mv^2$

where v is the velocity of the pendulum at this point.

Conservation of Energy

The total energy of the pendulum is the sum of its potential energy and kinetic energy:

$E = U + K$

a) At the extreme points, the total energy is:

   $E = mgh + 0 = mgh$

b) At the mean position, the total energy is:

   $E = 0 + (1/2)mv^2 = (1/2)mv^2$

c) At a point between the mean and extreme position, the total energy is:

   $E = mgh + (1/2)mv^2$

Since the total energy at all points is the same, we can conclude that energy is conserved in this case.

Mathematical Proof of Conservation of Energy

Let's consider the equation of motion of the pendulum:

$d^2\theta/dt^2 + (g/L)\sin\theta = 0$

where θ is the angle of displacement, g is the acceleration due to gravity, and L is the length of the pendulum.

Multiplying this equation by $d\theta/dt$, we get:

$(d\theta/dt)(d^2\theta/dt^2) + (g/L)\sin\theta\,(d\theta/dt) = 0$

Integrating this equation with respect to time, we get:

$(1/2)(d\theta/dt)^2 + (g/L)(1 - \cos\theta) = C$

where C is a constant.

The first term on the left-hand side represents the kinetic energy of the pendulum, while the second term represents the potential energy. Since the sum of these two terms is constant, we can conclude that energy is conserved.

Therefore, we have mathematically proven that energy is conserved in the case of a simple pendulum.

Power

Power is the rate at which work is done or energy is transferred. It is a measure of how quickly a task can be accomplished. In physics, power is defined as the rate of doing work or the rate of transfer of energy.

Mathematical Definition of Power

Power (P) is defined mathematically as:

$P = W / t$

where W is the work done and t is the time taken to do the work.

Alternative Definition of Power

Power can also be defined as:

$P = F \times v$

where F is the force applied and v is the velocity of the object.

Units of Power

Power is measured in various units, including:

1. Watt (W): The watt is the SI unit of power, and it is defined as one joule per second.
2. Horsepower (hp): The horsepower is a unit of power that is commonly used to measure the power of engines and motors. One horsepower is equal to 746 watts.
3. Kilowatt (kW): The kilowatt is a unit of power that is equal to one thousand watts.
4. Megawatt (MW): The megawatt is a unit of power that is equal to one million watts.
5. Gigawatt (GW): The gigawatt is a unit of power that is equal to one billion watts.

Relations Between Units of Power

Here are some relations between units of power:

1. Watt (W) and Horsepower (hp): 1 hp = 746 W
2. Watt (W) and Kilowatt (kW): 1 kW = 1000 W
3. Watt (W) and Megawatt (MW): 1 MW = 1,000,000 W
4. Watt (W) and Gigawatt (GW): 1 GW = 1,000,000,000 W
5. Horsepower (hp) and Kilowatt (kW): 1 hp = 0.7457 kW

6. Kilowatt (kW) and Megawatt (MW): 1 MW = 1000 kW
7. Megawatt (MW) and Gigawatt (GW): 1 GW = 1000 MW

Conversion Factors
Here are some conversion factors between units of power:

1. Watt (W) to Horsepower (hp): 1 W = 0.00134 hp
2. Horsepower (hp) to Watt (W): 1 hp = 746 W
3. Watt (W) to Kilowatt (kW): 1 W = 0.001 kW
4. Kilowatt (kW) to Watt (W): 1 kW = 1000 W
5. Watt (W) to Megawatt (MW): 1 W = 0.000001 MW
6. Megawatt (MW) to Watt (W): 1 MW = 1,000,000 W

In conclusion, power is measured in various units, including watt, horsepower, kilowatt, megawatt, and gigawatt. Understanding the relations between these units and using conversion factors can help to convert between them.

Relationship Between Power, Force, and Displacement

Power is related to force and displacement through the work-energy theorem, which states that the work done on an object is equal to the change in its kinetic energy. Mathematically, this can be expressed as:
$$W = \Delta K$$
where ΔK is the change in kinetic energy.
Since power is the rate of doing work, we can write:
$$P = W / t = \Delta K / t$$

Relationship Between Power and Energy
Power is also related to energy through the equation:

$P = \Delta E / t$
where ΔE is the change in energy.

Principles of Power

There are several principles of power that are important to understand:

1. **Power is a scalar quantity**: Power is a measure of the rate of energy transfer, and it has no direction.
2. **Power is a rate**: Power is a measure of the rate at which energy is transferred or work is done.
3. **Power is dependent on force and velocity**: The power delivered by a force is dependent on the magnitude of the force and the velocity of the object.

Uses and Applications of Power

Power has numerous uses and applications in various fields, including:

1. **Engineering**: Power is used to design and optimize systems, such as engines, motors, and generators.
2. **Physics**: Power is used to study the behavior of physical systems, such as the motion of objects and the transfer of energy.
3. **Biology**: Power is used to study the behavior of living organisms, such as the movement of muscles and the transfer of energy in cells.

Concepts and Conceptual Problems

Here are some concepts and conceptual problems related to power:

1. What is the difference between power and energy?
Answer: Power is the rate of energy transfer, while energy is the capacity to do work.
2. How is power related to force and velocity?
Answer: Power is equal to the product of force and

velocity.
3. What is the unit of power?
Answer: The unit of power is the watt (W).

Mathematical Operations
Here are some mathematical operations related to power:
1. Calculating power from work and time:
$P = W / t$
2. Calculating power from force and velocity:
$P = F \times v$
3. Calculating energy from power and time:
$E = P \times t$

Diagram
Here is a diagram showing the relationship between power, force, and velocity:
Force (F) → Velocity (v) → Power (P)

Examples
Here are some examples of power:
1. A car engine: The power of a car engine is measured in horsepower (hp) or watts (W).
2. A light bulb: The power of a light bulb is measured in watts (W).
3. A human muscle: The power of a human muscle is measured in watts (W) or horsepower (hp).

In conclusion, power is the rate at which work is done or energy is transferred. It is a measure of how quickly a task can be accomplished. Power is related to force and velocity, and it has numerous uses and applications in various fields.

In physics, systems can be classified into two main categories: closed systems and

open systems.

Closed System:
A closed system is a system that does not exchange matter with its surroundings, but can exchange energy. In other words, no mass can enter or leave the system, but energy can be transferred across the system boundary.
Example: A sealed container filled with gas, a thermos flask, or a car engine.

Open System:
An open system is a system that can exchange both matter and energy with its surroundings. In other words, mass and energy can both enter and leave the system.
Example: A river, a living organism, or a car engine with fuel intake and exhaust.

Energy Transfer in Closed Systems:
In a closed system, energy can be transferred in the form of heat, work, or radiation. The total energy of the system remains constant, but the form of energy can change. Mathematically, this can be represented by the equation:
$\Delta E = Q - W$
where ΔE is the change in energy, Q is the heat added to the system, and W is the work done by the system.

Energy Transfer in Open Systems:
In an open system, energy can be transferred in the form of heat, work, radiation, or mass transfer. The total energy of the system can change due to the exchange of matter and energy with the surroundings.

Mathematically, this can be represented by the equation:

$$\Delta E = Q - W + \Delta m \times h$$

where ΔE is the change in energy, Q is the heat added to the system, W is the work done by the system, Δm is the change in mass, and h is the specific enthalpy of the mass transferred.

Mathematical Operations:
1. Energy conservation: $\Delta E = 0$ (closed system)
2. Energy transfer: $\Delta E = Q - W$ (closed system)
3. Energy transfer with mass transfer: $\Delta E = Q - W + \Delta m \times h$ (open system)

Note: *These equations are simplified and assume a constant pressure and temperature. In real-world applications, additional factors like friction, heat transfer, and non-equilibrium processes may need to be considered.*

Here are some uses and applications of closed and open systems:

Uses and Applications of Closed Systems:

1. Thermos Flasks: Closed systems are used in thermos flasks to keep liquids at a constant temperature by reducing heat transfer.

2. Refrigeration Systems: Closed systems are used in refrigeration systems to transfer heat from one location to another.

3. Air Conditioning Systems: Closed systems are used in air conditioning systems to control temperature and humidity.

4. Internal Combustion Engines: Closed systems are used in internal combustion engines to convert chemical energy into mechanical energy.

5. Spacecraft: Closed systems are used in spacecraft to maintain a stable internal environment and conserve resources.

Uses and Applications of Open Systems:
1. Power Plants: Open systems are used in power plants to generate electricity by converting energy from one form to another.
2. Chemical Plants: Open systems are used in chemical plants to manufacture chemicals by converting raw materials into products.
3. Biological Systems: Open systems are used in biological systems, such as living organisms, to maintain homeostasis and respond to environmental changes.
4. Ecosystems: Open systems are used in ecosystems to describe the interactions between living organisms and their environment.
5. Agricultural Systems: Open systems are used in agricultural systems to describe the interactions between crops, soil, water, and air.

Real-World Examples:
1. Human Body: The human body is an open system that exchanges matter and energy with its environment to maintain homeostasis.
2. Car Engine: A car engine is a closed system that converts chemical energy into mechanical energy, but it is part of a larger open system that includes the car, the road, and the environment.
3. Power Grid: A power grid is an open system that generates, transmits, and distributes electricity to meet the energy demands of a region.

4. Ecosystem: An ecosystem is an open system that includes living organisms, soil, water, and air, and exchanges matter and energy with its environment.

5. Manufacturing Plant: A manufacturing plant is an open system that converts raw materials into products, and exchanges matter and energy with its environment.

Here are some real-world applications of work, energy, and power:

Work
1. Hydroelectric Power Plants: Water stored behind a dam has potential energy, which is converted into kinetic energy as it flows down the dam, generating electricity.
2. Internal Combustion Engines: Chemical energy stored in fuel is converted into kinetic energy, which powers the engine.
3. Cranes and Pulleys: Work is done by applying a force to lift or move heavy loads.
4. Bicycles: Pedaling a bicycle converts chemical energy from food into kinetic energy, propelling the bike forward.

Energy
1. Solar Panels: Light energy from the sun is converted into electrical energy, which is used to power homes and devices.
2. Wind Turbines: Kinetic energy from wind is converted into electrical energy, generating power.
3. Battery-Powered Devices: Chemical energy stored in batteries is converted into electrical energy, powering devices like smartphones and laptops.

4. Geothermal Power Plants: Heat energy from the Earth's core is converted into electrical energy, generating power.

Power
1. Electric Motors: Electrical energy is converted into mechanical energy, powering devices like fans and pumps.
2. Generators: Mechanical energy is converted into electrical energy, generating power for homes and industries.
3. Air Conditioners: Electrical energy is converted into thermal energy, cooling buildings and homes.
4. Rocket Propulsion: Chemical energy stored in fuel is converted into kinetic energy, propelling rockets into space.

These are just a few examples of the many real-world applications of work, energy, and power.

Here are twenty "what happens when" questions with answers based on work, energy, and power from Class XI Physics:

1. What happens when a force is applied to an object in the direction of its motion?
Answer: The object's kinetic energy increases.
2. What happens when a ball is thrown upwards?
Answer: Its kinetic energy decreases and potential energy increases.
3. What happens when a car accelerates from rest?
Answer: Its kinetic energy increases.
4. What happens when a bicycle is pedaled?
Answer: Chemical energy from the rider's body is converted into kinetic energy.
5. What happens when a weight is lifted to a certain

height?
Answer: Its potential energy increases.
6. What happens when a spring is compressed?
Answer: Its potential energy increases.
7. What happens when a ball is rolled on a rough surface?
Answer: Its kinetic energy decreases due to friction.
8. What happens when a car brakes?
Answer: Its kinetic energy is converted into heat energy.
9. What happens when a generator is used to produce electricity?
Answer: Mechanical energy is converted into electrical energy.
10. What happens when a battery is connected to a circuit?
Answer: Chemical energy is converted into electrical energy.
11. What happens when a person climbs a staircase?
Answer: Chemical energy from the person's body is converted into potential energy.
12. What happens when a skier slides down a hill?
Answer: Potential energy is converted into kinetic energy.
13. What happens when a car is driven uphill?
Answer: Kinetic energy is converted into potential energy.
14. What happens when a rubber band is stretched?
Answer: Its potential energy increases.
15. What happens when a toy car is wound up?
Answer: Potential energy is stored in the spring.

16. What happens when a person jumps from a height?

Answer: Potential energy is converted into kinetic energy.

17. What happens when a bicycle is coasting downhill?
Answer: Potential energy is converted into kinetic energy.

18. What happens when a weight is dropped from a height?
Answer: Potential energy is converted into kinetic energy.

19. What happens when a spring is released from compression?
Answer: Potential energy is converted into kinetic energy.

20. What happens when a car's brakes are applied on a slope?
Answer: Kinetic energy is converted into heat energy, and potential energy increases.

Here are ten solved numerical problems

Problem 1

A force of 10 N acts on an object and displaces it by 5 m. Calculate the work done.

Step 1: Identify the given values

Force (F) = 10 N, Displacement (s) = 5 m

Step 2: Use the formula for work done

Work done (W) = F × s = 10 N × 5 m = 50 J

The final answer is: 50

Problem 2

A body of mass 20 kg is lifted to a height of 10 m. Calculate the potential energy stored in the body. (g

$= 9.8$ m/s²)

Step 1: Identify thé given values

Mass (m) = 20 kg, Height (h) = 10 m, Acceleration due to gravity (g) = 9.8 m/s²

Step 2: Use the formula for potential energy

Potential energy (U) = mgh = 20 kg × 9.8 m/s² × 10 m = 1960 J

The final answer is: 1960

Problem 3

A car accelerates from rest to a velocity of 20 m/s in 5 s. Calculate the power developed by the car's engine.

Step 1: Identify the given values

Initial velocity (u) = 0 m/s, Final velocity (v) = 20 m/s, Time (t) = 5 s

Step 2: Calculate the acceleration

Acceleration (a) = $\Delta v / \Delta t$ = (20 m/s - 0 m/s) / 5 s = 4 m/s²

Step 3: Calculate the power

Power (P) = Force × velocity = mass × acceleration × velocity (assuming F = ma)

P = 4 m/s² × 20 m/s = 80 W (assuming a mass of 1 kg for simplicity)

The final answer is: 80

Problem 4

A 50 kg person climbs a staircase of height 10 m in 5 s. Calculate the power developed by the person.

Step 1: Identify the given values

Mass (m) = 50 kg, Height (h) = 10 m, Time (t) = 5 s

Step 2: Calculate the work done

Work done (W) = mgh = 50 kg × 9.8 m/s² × 10 m = 4900 J

Step 3: Calculate the power

Power (P) = Work done / Time = 4900 J / 5 s = 980 W

The final answer is: 980

Problem 5

A block of mass 10 kg is pulled by a force of 20 N through a distance of 5 m. Calculate the work done and the increase in kinetic energy.

Step 1: Identify the given values

Mass (m) = 10 kg, Force (F) = 20 N, Distance (s) = 5 m

Step 2: Calculate the work done

Work done (W) = F × s = 20 N × 5 m = 100 J

Step 3: Calculate the increase in kinetic energy

Increase in kinetic energy = Work done = 100 J

The final answer is: 100

Problem 6

A car of mass 1500 kg accelerates from rest to a velocity of 30 m/s in 10 s. Calculate the power developed by the car's engine.

Step 1: Identify the given values

Mass (m) = 1500 kg, Initial velocity (u) = 0 m/s, Final velocity (v) = 30 m/s, Time (t) = 10 s

Step 2: Calculate the acceleration

Acceleration (a) = Δv / Δt = (30 m/s - 0 m/s) / 10 s = 3 m/s²

Step 3: Calculate the power

Power (P) = Force × velocity = mass × acceleration × velocity (assuming F = ma)

P = 3 m/s² × 30 m/s × 1500 kg = 135000 W

The final answer is: 135000

Problem 7

A body of mass 20 kg is moving with a velocity of 10 m/s. Calculate the kinetic energy of the body.

Step 1: Identify the given values

Mass (m) = 20 kg, Velocity (v) = 10 m/s

Step 2: Calculate the kinetic energy

Kinetic energy (K) = $(1/2)mv^2$ = (1/2) × 20 kg × (10 m/s)² = 100

Problem 8

A 2 kW electric motor is used to lift a load of 500 kg to a height of 10 m in 5 s. Calculate the efficiency of the motor.

Step 1: Identify the given values

Power (P) = 2 kW = 2000 W, Mass (m) = 500 kg, Height (h) = 10 m, Time (t) = 5 s

Step 2: Calculate the work done

Work done (W) = mgh = 500 kg × 9.8 m/s² × 10 m = 49000 J

Step 3: Calculate the energy input

Energy input (E) = Power × Time = 2000 W × 5 s = 10000 J

Step 4: Calculate the efficiency

Efficiency (η) = Work done / Energy input = 49000 J / 10000 J = 0.49 or 49%

The final answer is: 49%

Problem 9

A car of mass 1000 kg accelerates from rest to a velocity of 20 m/s in 10 s. Calculate the average power developed by the car's engine.

Step 1: Identify the given values

Mass (m) = 1000 kg, Initial velocity (u) = 0 m/s, Final velocity (v) = 20 m/s, Time (t) = 10 s

Step 2: Calculate the acceleration

Acceleration (a) = $\Delta v / \Delta t$ = (20 m/s - 0 m/s) / 10 s = 2 m/s²

Step 3: Calculate the force

Force (F) = mass × acceleration = 1000 kg × 2 m/s² = 2000 N

Step 4: Calculate the average power

Average power (P) = Force × velocity / Time = 2000 N × 20 m/s / 10 s = 4000 W

The final answer is: 4000

Problem 10

A block of mass 5 kg is moving with a velocity of 10 m/s on a frictionless surface. Calculate the kinetic energy of the block.

Step 1: Identify the given values

Mass (m) = 5 kg, Velocity (v) = 10 m/s

Step 2: Calculate the kinetic energy

Kinetic energy (K) = $(1/2)mv^2$ = $(1/2) \times 5$ kg $\times (10$ m/s$)^2$ = 250 J

The final answer is: 250

1. A force of 10 N acts on an object and displaces it by 5 m. Calculate the work done.

Answer: 50 J

Solve:

1. A body of mass 20 kg is lifted to a height of 10 m. Calculate the work done.

Answer: 1960 J

2. A force of 50 N acts on an object and displaces it by 2 m. Calculate the work done.

Answer: 100 J

3. A body of mass 10 kg is moving with a velocity of 5 m/s. Calculate its kinetic energy.

Answer: 125 J

4. A body of mass 20 kg is lifted to a height of 10 m. Calculate its potential energy.

Answer: 1960 J

5. A car of mass 1500 kg is moving with a velocity of 30 m/s. Calculate its kinetic energy.

Answer: 675000 J

6. A force of 10 N acts on an object and displaces it by 5 m in 2 s. Calculate the power.

Answer: 25 W

7. A body of mass 20 kg is lifted to a height of 10 m in 5 s. Calculate the power.

Answer: 392 W

8. A car of mass 1500 kg is moving with a velocity of 30 m/s. Calculate the power of the car's engine if it accelerates from rest to this velocity in 10 s.

Answer: 67500 W

Here are ten difficult numerical questions based on work, energy, and power with solutions:

Problem 1

A block of mass 10 kg is pulled by a force of 50 N through a distance of 20 m on a rough surface. If the coefficient of friction between the block and the surface is 0.2, calculate the work done by the force.

Solution

Force (F) = 50 N, Distance (s) = 20 m, Coefficient of friction (μ) = 0.2

Work done (W) = F × s - μ × m × g × s = 50 N × 20 m - 0.2 × 10 kg × 9.8 m/s² × 20 m = 1000 J - 392 J = 608 J

Problem 2

A car of mass 1500 kg accelerates from rest to a velocity of 60 m/s in 10 s. Calculate the power developed by the car's engine.

Solution

Mass (m) = 1500 kg, Initial velocity (u) = 0 m/s, Final velocity (v) = 60 m/s, Time (t) = 10 s

Acceleration (a) = $\Delta v / \Delta t$ = (60 m/s - 0 m/s) / 10 s = 6 m/s²

Force (F) = mass × acceleration = 1500 kg × 6 m/s² = 9000 N

Power (P) = Force × velocity / Time = 9000 N × 60

m/s / 10 s = 540000 W

Problem 3

A body of mass 20 kg is moving with a velocity of 10 m/s. Calculate its kinetic energy. If the body is brought to rest by a force of 50 N, calculate the distance over which the force acts.

Solution

Mass (m) = 20 kg, Velocity (v) = 10 m/s

Kinetic energy (K) = $(1/2)mv^2$ = $(1/2) \times 20$ kg $\times (10$ m/s$)^2$ = 1000 J

Force (F) = 50 N, Kinetic energy (K) = 1000 J

Distance (s) = Kinetic energy / Force = 1000 J / 50 N = 20 m

Problem 4

A spring of force constant 100 N/m is compressed by 20 cm. Calculate the work done by the spring force.

Solution

Force constant (k) = 100 N/m, Compression (x) = 20 cm = 0.2 m

Work done (W) = $(1/2)kx^2$ = $(1/2) \times 100$ N/m $\times (0.2$ m$)^2$ = 2 J

Problem 5

A car of mass 1500 kg accelerates from rest to a velocity of 30 m/s in 10 s. Calculate the power developed by the car's engine. If the car's engine is 25% efficient, calculate the energy input to the engine.

Solution

Mass (m) = 1500 kg, Initial velocity (u) = 0 m/s, Final velocity (v) = 30 m/s, Time (t) = 10 s

Acceleration (a) = $\Delta v / \Delta t$ = (30 m/s - 0 m/s) / 10 s = 3 m/s²

Force (F) = mass × acceleration = 1500 kg × 3 m/s² = 4500 N

Power (P) = Force × velocity / Time = 4500 N × 30 m/s / 10 s = 135000 W

Efficiency (η) = 25% = 0.25, Power (P) = 135000 W

Energy input (E) = Power / Efficiency = 135000 W / 0.25 = 540000 J

Problem 6

A body of mass 10 kg is moving with a velocity of 20 m/s. Calculate its kinetic energy. If the body is brought to rest by a force of 100 N, calculate the distance over which the force acts.

Solution

Mass (m) = 10 kg, Velocity (v) = 20 m/s

Kinetic energy (K) = (1/2)mv² = (1/2) × 10 kg × (20 m/s)² = 2000 J

Force (F) = 100 N, Kinetic energy (K) = 2000 J

Distance (s) = Kinetic energy / Force = 2000 J / 100 N = 20 m

Problem 7

A car of mass 2000 kg accelerates from rest to a velocity of 40 m/s in 15 s. Calculate the power developed by the car's engine. If the car's engine is 30% efficient, calculate the energy input to the engine.

Solution

Mass (m) = 2000 kg, Initial velocity (u) = 0 m/s, Final velocity (v) = 40 m/s, Time (t) = 15 s

Acceleration (a) = $\Delta v / \Delta t$ = (40 m/s - 0 m/s) / 15 s = 2.67 m/s²

Force (F) = mass × acceleration = 2000 kg × 2.67 m/s² = 5333 N

Power (P) = Force × velocity / Time = 5333 N × 40 m/s / 15 s = 142222 W

Efficiency (η) = 30% = 0.3, Power (P) = 142222 W

Energy input (E) = Power / Efficiency = 142222 W / 0.3 = 474074 J

Problem 8

A body of mass 50 kg is moving with a velocity of 15 m/s. Calculate its kinetic energy. If the body is brought to rest by a force of 200 N, calculate the distance over which the force acts.

Solution

Mass (m) = 50 kg, Velocity (v) = 15 m/s

Kinetic energy (K) = $(1/2)mv^2$ = (1/2) × 50 kg × (15 m/s)² = 5625 J

Force (F) = 200 N, Kinetic energy (K) = 5625 J

Distance (s) = Kinetic energy / Force = 5625 J / 200 N = 28.125 m

Problem 9

A spring of force constant 200 N/m is compressed by 10 cm. Calculate the work done by the spring force. If the spring is then released, calculate the maximum velocity of the attached mass.

Solution

Force constant (k) = 200 N/m, Compression (x) = 10 cm = 0.1 m

Work done (W) = $(1/2)kx^2$ = (1/2) × 200 N/m × (0.1 m)² = 1 J

Maximum velocity (v) = √(2 × Work done / Mass) = √(2 × 1 J / 0.1 kg) = 4.47 m/s

Problem 10

A car of mass 1500 kg accelerates from rest to a velocity of 25 m/s in 10 s. Calculate the power developed by the car's engine. If the car's engine is 25% efficient, calculate the energy input to the engine.

Solution

Mass (m) = 1500 kg, Initial velocity (u) = 0 m/s, Final velocity (v) = 25 m/s, Time (t) = 10 s

Acceleration (a) = $\Delta v / \Delta t$ = (25 m/s - 0 m/s) / 10 s = 2.5 m/s²

Force (F) = mass × acceleration = 1500 kg × 2.5 m/s² = 3750 N

Power (P) = Force × velocity / Time = 3750 N × 25 m/s / 10 s = 93750 W

Efficiency (η) = 25% = 0.25, Power (P) = 93750 W

Energy input (E) = Power / Efficiency = 93750 W / 0.25 = 375000 J

Here are the exercises with hints:

Very Short Answer Questions (1 mark each)

1. Define work done.

Hint: Work done is a measure of the energy transferred by a force to an object.

2. What is the SI unit of energy?

Hint: The SI unit of energy is Joule (J).

3. Define power.

Hint: Power is the rate at which work is done or energy is transferred.

4. What is the relationship between work done and energy transferred?

Hint: Work done is equal to the energy transferred.

5. Define kinetic energy.

Hint: Kinetic energy is the energy an object possesses due to its motion.

6. What is the relationship between potential energy and height?

Hint: Potential energy is directly proportional to height.

7. Define efficiency.

Hint: Efficiency is the ratio of output energy to input energy.

8. What is the SI unit of power?

Hint: The SI unit of power is Watt (W).

9. Define conservative forces.

Hint: Conservative forces are forces that conserve energy.

10. What is the relationship between work done and force applied?

Hint: Work done is equal to the product of force

applied and displacement.

Short Answer Questions (2-3 marks each)

1. Derive the expression for work done by a constant force.

Hint: Use the definition of work done and the equation of motion.

2. Explain the concept of kinetic energy.

Hint: Use examples to illustrate the concept.

3. Derive the expression for potential energy.

Hint: Use the concept of work done and the equation of motion.

4. Explain the concept of power.

Hint: Use examples to illustrate the concept.

5. Derive the expression for efficiency.

Hint: Use the definition of efficiency and the equation of motion.

6. Explain the concept of conservative forces.

Hint: Use examples to illustrate the concept.

7. Derive the expression for work-energy theorem.

Hint: Use the definition of work done and the equation of motion.

8. Explain the concept of potential energy and its relationship with height.

Hint: Use examples to illustrate the concept.

9. Derive the expression for kinetic energy in terms of momentum.

Hint: Use the definition of kinetic energy and the equation of motion.

10. Explain the concept of power and its relationship with work done.

Hint: Use examples to illustrate the concept.

Long Answer Questions (5-6 marks each)

1. Derive the expression for work done by a variable force.

Hint: Use the definition of work done and the equation of motion.

2. Explain the concept of energy and its different forms.

Hint: Use examples to illustrate the concept.

3. Derive the expression for kinetic energy in terms of velocity.

Hint: Use the definition of kinetic energy and the equation of motion.

4. Explain the concept of potential energy and its relationship with height.

Hint: Use examples to illustrate the concept.

5. Derive the expression for power in terms of work done and time.

Hint: Use the definition of power and the equation of motion.

6. Explain the concept of conservative forces and non-conservative forces.

Hint: Use examples to illustrate the concept.

7. Derive the expression for work-energy theorem in

terms of kinetic energy and potential energy.

Hint: Use the definition of work done and the equation of motion.

8. Explain the concept of efficiency and its relationship with energy transfer.

Hint: Use examples to illustrate the concept.

9. Derive the expression for kinetic energy in terms of momentum and velocity.

Hint: Use the definition of kinetic energy and the equation of motion.

10. Explain the concept of power and its relationship with work done and time.

Hint: Use examples to illustrate the concept.

Numerical Questions (3-4 marks each)

1. A force of 10 N acts on an object and displaces it by 5 m. Calculate the work done.

Hint: Use the definition of work done.

2. A body of mass 20 kg is moving with a velocity of 10 m/s. Calculate its kinetic energy.

Hint: Use the definition of kinetic energy.

3. A spring of force constant 100 N/m is compressed by 20 cm. Calculate the work done by the spring force.

Hint: Use the definition of work done.

4. A car of mass 1500 kg accelerates from rest to a velocity of 30 m/s in 10 s. Calculate the power developed by the car's engine.

Hint: Use the definition of power.

5. A body of mass 10 kg is moving with a velocity of 20 m/s. Calculate its kinetic energy. If the body is brought to rest by a force of 100 N, calculate the distance over which the force acts.

Hint: Use the definition of kinetic energy and work done.

6. A spring of force constant 50 N/m is stretched by 30 cm. Calculate the work done by the spring force.

Hint: Use the definition of work done.

7. A body of mass 30 kg is lifted to a height of 10 m. Calculate the potential energy of the body. If the body is dropped from this height, calculate its kinetic energy just before it hits the ground.

Hint: Use the definition of potential energy and kinetic energy.

8. A car of mass 1500 kg accelerates from rest to a velocity of 40 m/s in 15 s. Calculate the power developed by the car's engine. If the car's engine is 25% efficient, calculate the energy input to the engine.

Hint: Use the definition of power and efficiency.

9. A spring of force constant 200 N/m is compressed by 15 cm. Calculate the work done by the spring force. If the spring is then released, calculate the maximum velocity of the attached mass.

Hint: Use the definition of work done and kinetic energy.

10. A body of mass 20 kg is moving with a velocity of 15 m/s. Calculate its kinetic energy. If the body is

brought to rest by a force of 150 N, calculate the distance over which the force acts.

Hint: Use the definition of kinetic energy and work done.

Here is the table with serial numbers:

Device/Activity	Initial Energy Form	Converted Energy Form	process
1. Electric Light Bulb	Electrical Energy	Light Energy (Radiant Energy)	Electrical current heats up the filament, producing light
2. Car Engine	Chemical Energy (Fuel)	Mechanical Energy (Kinetic Energy)	propelling the car
3. Solar Panel	Radiant Energy (Solar Energy)	Electrical Energy	Photovoltaic cells convert sunlight into electrical energy
4. Human Body	Chemical Energy (Food)	Mechanical Energy (Kinetic Energy)	Digestion of food produces energy, which is converted into mechanical energy for movement
5. Refrigerator | Electrical Energy	Thermal Energy (Cooling)	Electrical energy powers the compressor	cooling the refrigerant and transferring heat from the interior to the exterior
6. Wind Turbine	Kinetic Energy (Wind)	Electrical Energy	Blades convert wind kinetic energy into

			mechanical energy, which is then converted into electrical energy
7. Battery Charger	|Electrical Energy	Chemical Energy (Stored Energy)	Electrical energy is converted into chemical energy, storing energy in the battery
8. Water Pump	Electrical Energy	Mechanical Energy (Kinetic Energy)	Electrical energy powers the motor, pumping water and converting electrical energy into mechanical energy
9. Microwave Oven	Electrical Energy	Thermal Energy (Heat)	Electrical energy produces microwave radiation, which is absorbed by food, producing heat
10. Bicycle	Chemical Energy (Food)	Mechanical Energy (Kinetic Energy)	Pedaling converts chemical energy from food into mechanical energy, propelling the bicycle

Here are some interesting facts about the chapter on Work, Energy, and Power:

1. Energy is everywhere: Energy is present in all forms of matter, from the food we eat to the stars in the sky.

2. Work and energy are interchangeable: According to the work-energy theorem, the work done on an object is equal to the change in its kinetic energy.

3. Power is the rate of energy transfer: Power is measured in watts (W) and represents the rate at which energy is transferred or converted from one form to another.

4. The law of conservation of energy is universal: The law states that energy cannot be created or destroyed, only converted from one form to another. This law applies to all forms of energy, from the smallest subatomic particles to the entire universe.

5. Vectors are essential in physics: Vectors are used to represent forces, velocities, and accelerations in physics. They have both magnitude and direction, making them essential for describing complex phenomena.

6. The parallelogram law of vectors is a fundamental concept: This law states that when two vectors are added, the resultant vector is the diagonal of a parallelogram formed by the two vectors.

7. Energy conversion is a crucial aspect of engineering: Engineers design systems to convert energy from one form to another, such as from electrical energy to mechanical energy in motors.

Here are some scientists who made significant contributions to the concepts discussed in this chapter:

1. **Sir Isaac Newton** (1643-1727): Newton laid the foundation for classical mechanics, including the concepts of work, energy, and power. His laws of motion and universal gravitation are still widely used today.

2. **Antoine Lavoisier** (1743-1794): Lavoisier discovered the concept of energy conservation and recognized that energy cannot be created or destroyed, only converted from one form to another.

3. **James Joule** (1818-1889): Joule discovered the mechanical equivalent of heat, which led to the development of the concept of energy. He also formulated the law of conservation of energy.

4. **Hermann von Helmholtz** (1811-1894): Helmholtz formulated the law of conservation of energy and recognized that energy is a fundamental physical quantity.

5. **William Thomson (Lord Kelvin)** (1824-1907): Thomson developed the concept of energy and its relationship to work and heat. He also formulated the second law of thermodynamics.

6. **Alessandro Volta (1745-1827)**: Volta invented the electric battery, which led to the development of electrical energy and its relationship to work and power.

7. **Michael Faraday (1791-1867):** Faraday discovered the principles of electromagnetic induction, which led to the development of electrical generators and motors.

8. **Gottfried Wilhelm Leibniz (1646-1716):**

Leibniz developed the concept of vis viva (living force), which is equivalent to kinetic energy. He also formulated the law of conservation of energy.

www.ingramcontent.com/pod-product-compliance
Lightning Source LLC
Chambersburg PA
CBHW031617210526
45464CB00004B/1623